Science Writing for Beginners

BY THE SAME AUTHOR

Textbooks

A Laboratory Handbook of Blood Transfusion Techniques

A Synopsis of Blood Grouping Theory and Serological Techniques

History of Institutions

Let Not the Deep: the story of the
Royal National Lifeboat Institution

Learn, that You may Improve: the history of the
Institute of Medical Laboratory Sciences

Transport History

The Royal Deeside Line

Stories of Royal Deeside's Railway

The Campbeltown and Machrihanish Light Railway

Philosophy

God, Blood and Society

Science Writing for Beginners

A.D. FARR
BA, PhD, FIMLS
Editor, *Medical Laboratory Sciences*

BLACKWELL SCIENTIFIC PUBLICATIONS
OXFORD LONDON EDINBURGH
BOSTON PALO ALTO MELBOURNE

Blackwell Scientific Publications
Editorial offices:
Osney Mead, Oxford, OX2 0EL
8 John Street, London, WC1N 2ES
23 Ainslie Place, Edinburgh, EH3 6AJ
52 Beacon Street, Boston
Massachusetts 02108, USA
744 Cowper Street, Palo Alto
California 94301, USA
107 Barry Street, Carlton
Victoria 3053, Australia

First published 1985

Photoset by Enset (Photosetting)
Midsomer Norton, Bath, Avon
and printed and bound in
Great Britain by
Biddles Ltd, Guildford

DISTRIBUTORS

USA and Canada
Blackwell Scientific Publications Inc
PO Box 50009, Palo Alto
California 94303

Australia
Blackwell Scientific Book Distributors
31 Advantage Road, Highett
Victoria 3190

British Library
Cataloguing in Publication Data

Farr, A.D.
Science writing for beginners.
1. Technical writing
I. Title
808′.0665021 T11

ISBN 0–632–01362–1

'Learn to write well, or not to write at all.'

John Dryden
Essay on Satire, 1693

Contents

List of Illustrations

Introduction
Why must scientists write?

> Reading maketh a full man; conference a ready man; and writing an exact man.
>
> Francis Bacon (1561–1624)
> *Of studies*

'To be able to write clear and simple English is perhaps the most generally useful of all educational skills. Whatever your profession, you will need at some time to write reports, to present technical information or to compile statements and memoranda. The influence that you have in the world will depend very much on your ability to put your thoughts on paper. More immediately, in writing essays and examination papers, you will need to be able to write clearly and intelligibly'[1].

Now although Maddox was writing about the general principles of *How to Study* these words of his sum up very well why scientists in particular need to be able to write coherently and effectively. Science is primarily a practical occupation, but it is not sufficient to be able to 'do' science if one cannot 'write' science also[2]. From the simple business of writing reports of particular analyses to the production of theses and text books, during any scientist's working life there is a constant need to communicate. Yet a century ago Sir James Barrie commented drily that 'the scientific man is the only person who has anything new to say and who does not know how to say it.'[3] The most carefully and accurately performed laboratory test is worthless if the investigator is unable to communicate the result in a form which others can understand. In the pressures of modern life it is often averred that there is another reason why scientists must write: it is summed up in the maxim 'publish or perish'. Probably too often, when a scientist applies for a new post or for a research grant, one of the principal criteria used to assess the applicant's suitability is the number of 'publications' listed in his *curriculum vitae*. At one time this was a not unreasonable idea, as the papers he or she had published were a fair index of the scientist's research activities. Today, however, the tail is wagging the dog and scientists feel impelled to publish something—anything—in order to keep up with the rat race. This is a pity but it should not let you lose sight of the fact that publications which have got past the process of peer review—assessment by fellow scientists—and been found worthy of publication really are a measure of your research or developmental capabilities.

At a more basic level, before anyone can be called a scientist it is necessary for him or her to obtain a recognized qualification, which in turn necessitates passing examinations—and it does not matter how much knowledge or skill one has acquired if one is unable to demonstrate to the examiner the extent of that learning.

In the more leisurely days prior to the middle of the present century, education was essentially a broadening experience in which everyone was taught 'the arts' and a few then went on to study 'the sciences'. All Oxford and Cambridge medical students took an arts degree before proceeding to their medical studies and every university required a competence in English, a classical language and often enough a second modern language, as part of their general entrance requirements. The overall effect was that while there were few scientists they were nearly all literate: today the pendulum has swung in the opposite direction and now there are many scientists, few of whom have ever been taught to take the use of language seriously, let alone to consider it a skill which they must acquire as part of their professional education.

All of this does not mean that every scientist has to become an expert in the finer points of grammatical synthesis and analysis, able to quote at the drop of a hat the precise definition of a dangling participle or to cite an example of an ablative absolute. To communicate clearly does not require a technical mastery of grammatical niceties so much as a grasp of the mechanics of constructing a concise and coherent report. Put another way: it is not so much what one has to do as what one must avoid doing that is important to the scientist who would write well.

Whenever one writes it is always for the benefit of a reader—who might be a colleague, an examiner, or even a critic—and it is always necessary to remember the needs of that reader. One must write clearly, not obscuring the message with 'waffle' or 'padding' and not leaving one's readers to puzzle out for themselves the sense of what one is saying. Provision of an informative title and effective headings and sub-headings; presentation of information in a logical order, including all the steps of an argument; clearly illustrated evidence and examples: these are the essentials of scientific writing and they are not difficult requirements to meet.

Anyone who has taken this book into his hands and read this far has taken the first (and most vital) step in acquiring a facility in science writing—they have realized that writing is an art which can, and should, be learned. It is not difficult to develop the necessary skills. Any scientist worth his salt already possesses the essential attributes of being capable of bringing a logical approach to a new problem and the ability to be self-critical: the only other factor needed is recognition of the nature of the problem. So if you are still with me and are sitting comfortably, let us begin.

Chapter 1
Writing good English

> This is the sort of English up with which I will not put.
>
> Winston Churchill (1874–1965)
> *Marginal note on document*

> True ease in writing comes from art, not chance.
>
> Alexander Pope (1688–1744)
> *An essay on criticism*

Before you turn rapidly to some other part of this book let me make it clear that this chapter is not a detailed study of the technicalities of grammar. It is simply a beginner's guide to the basics of what to do and what not to do.

When you write anything intended to be read by someone else—be it a report, an examination paper or even a personal letter—you do so with the intention of being understood. If you write poor English which does not clearly and logically express what you wish to say, which contains poor syntax or wrong spellings, or which is carelessly (usually excessively) punctuated your reader, or examiner, will form a poor opinion of you and will be left with the impression that you are uneducated or badly taught or—what is worse—that you don't understand yourself what you are trying to say. As Winston Churchill observed in an election speech in 1906, 'Men will forgive anything except bad prose'.[1]

The 'rules' of grammar exist not because grammarians have built up a defensive wall of jargon to keep others out, but to ensure that language does its job of conveying thoughts and ideas efficiently. Just as there have to be rules and signposts governing use of the roads by motor traffic in order to prevent chaos, so there are rules and signposts to prevent chaos in communication. Instead of one-way streets we have syntax (the order in which words must appear if the sentence is not to lose its meaning), and instead of 'Slow' and 'Halt' signs we have commas and full stops. If you can learn the

Highway Code you can learn the rules of grammar. However, writing—like driving—is a skill which has to be learned: it very rarely comes completely naturally. Don't make the mistake of considering the correct use of English too unimportant to bother with. Language is the outward expression of people's thoughts and to misuse the one is to misunderstand the other.

Stated simply there are five stages in any written composition:

1 You have ideas (or facts) which you wish to pass on to others.

2 These ideas have to be put into words and phrases and written down.

3 The words and phrases must run together into sentences which are grammatically correct.

4 The sentences must follow one another in a logical order, separated into units of thinking called paragraphs.

5 The final result must have the impact which you intended upon the reader.

Choice of words

Most people tend to find themselves torn in sympathy between Winnie-the-Pooh, who said that he was 'a Bear of Very Little Brain, and long words bother me'[2], and Humpty Dumpty. '"When I use a word", Humpty Dumpty said in rather a scornful tone, "it means just what I choose it to mean—neither more nor less"'.[3] The basic rules are to use the words which first come into your mind (always avoiding slang and made-up words not to be found in a dictionary), and to use short and familiar words rather than those which are long and unfamiliar. Never use six words where one will do. 'In view of the fact that' is much better expressed by 'As'; 'in the contemplated eventuality' is more clearly phrased as 'if so'. Similarly, Anglo-Saxon words are generally better than foreign equivalents. It is better to say 'see above' than to use the time-honoured (and time-worn) *vide supra,* and more people will understand 'delusions of greatness' than *folie de grandeur*. This is not to say that one must always strenuously avoid long or foreign words. Some Latin and French words, or abbreviations of them, have become almost English by adoption. Everyone understands *via* even if they have never studied Latin, and there is simply no short way of expressing long scientific words like 'chromatogram', 'electrophoresis' or 'photosynthesis'.

If you must use foreign words be very sure of their meaning and their spelling, and if you are writing for publication underline them to indicate that they should be printed in italics. Don't be tempted to mix foreign and English words in the same phrase. Some of the commoner foreign words you may wish to use are:

a priori = from cause to effect
ab initio = from the beginning
ad hoc = for this particular purpose
de facto = in fact, actually
de novo = newly, again
in situ = in position
in toto = entirely, completely
inter alia = amongst other things
per annum = each year
per diem = daily
per se = of itself
rationale = reasoning

Do watch out for plural forms of foreign words, however. Four examples of commonly misused words are:

Singular	*Plural*
Criterion	Criteria
Datum	Data (NB, 'the data are . . .')
Medium	Media
Phenomenon	Phenomena

Also beware of abbreviations. Some initial letters (like ml for millilitre, or DNA for deoxyribonucleic acid) are so much in everyday use that there is no problem, but unless there is an international convention for a particular abbreviation you may only confuse your reader. 'CPD' may mean either 'citrate-phosphate-dextrose solution' or 'congenital pelvic deformity' depending upon whether your reader is primarily interested in blood transfusion or gynaecology. If he is interested in both he may well become confused. Less obviously, there is a risk in abbreviating words by using only their first part. Long words are frequently compound adjectives and to separate them only confuses the meaning. For example, 'hypodermic' and 'hypoglycaemic' are both compound words each composed of two different adjectives: to say merely 'hypo' is to beg the question 'hypo-what?' It also gives no clue as to what noun or verb is intended to follow the compound word. Does one mean hypodermic syringe, or injection, or location? Or does one even mean 'hypodermic' at all? Leaving your reader to puzzle out your meaning is neither good use of English nor very helpful to the clear transmission of ideas.

Bringing together my last two points, the abbreviation of Latin expressions is fairly common—indeed it tends to be overdone—and some confusion often arises about just what particular letters mean. You should not really need to use more than those listed below. If you do then you would probably be better saying it in plain English.

cf.	= *confer*	= compare
et al.	= *et alii*	= and other people
etc.	= *et cetera*	= and the rest, and so forth
e.g.	= *exempli gratia*	= for example
i.e.	= *id est*	= that is
NB	= *nota bene*	= note well
per cent	= *per centum*	= by the hundred
q.v.	= *quod vide*	= which see

Most people tend to confuse e.g. with i.e. They do not mean the same, as you can see.

One of the commonest problems in writing is avoiding jargon and clichés. These are mostly perfectly good words (and phrases) which have become spoiled by careless and excessive use until they no longer mean very much; and often, indeed, they come to be so commonly misunderstood that any meaning they do retain is quite changed from that to be found in a dictionary. Such words as 'formulate', 'pinpoint' and 'crucial' should normally be avoided. Because such words are used excessively they may be those which you think of first, but beware of the dangers. Used as an adjective 'grass roots' (opinion) may well describe the lowest form of vegetable life you can think of, but that is probably not what you were trying to say.

Another common problem is the modern tendencey to make up words rather than use those already in existence. In particular there is a tendency to proliferate verbal nouns (and even noun-verbs: I recently heard someone say 'give it an invert' when she meant 'invert it'), and these often make nonsense when one considers what they really mean. For example, a frequent mistake made by biomedical scientists is with the noun 'aliquot'. An aliquot is one of a number of equal portions into which a volume of fluid may be separated; yet editors and examiners often read descriptions of how (e.g.) a scientist 'aliquoted' a serum in 200 μl amounts. Unless the original volume was exactly divisible by 200 they could not be aliquots; and what is really meant is 'dispensed' which is a perfectly good verb while 'to aliquot' is not—at least, not in British English. Such misuse of words is not only misleading but is ugly and only serves to add to the reputation of scientists as uneducated (even if well trained) and only semi-literate. Or maybe the anonymous diplomat quoted by Sir Ernest Gowers was right when he observed that 'what appears to be sloppy or meaningless use of words may well be a completely correct use of words to express sloppy or meaningless ideas'.[4]

Construction of sentences

Any word standing on its own has little meaning, even if it may seem obvious at first glance. For example, did you know that the *Oxford English Dictionary* gives 91 meanings for the word 'take'? So the meaning of a word must depend on its context—and this means not only the other words which surround it but also the order in which they appear.

Generally speaking a sentence is a group of words which is complete in itself and should contain only a single assertion, exclamation or comment (e.g. 'The tests should be incubated at 37°C.'; 'These results were unexpected!); although a sentence may sometimes contain more than one question or command (e.g. 'What is your name and your address?'; 'Add the di-ethyl ether and then agitate the tube.'). Put another way, within such sentences there will be one or more clauses—that is, groups of words each of which could stand on its own. Generally speaking a clause will contain a subject followed by a verb, and often an object and a complement, thus:

Subject (a noun)	Verb	Object (another noun)	Complement (adjective or adverb)
The physicist	labelled	the compound	correctly

Now if all this seems to be treading deep grammatical waters we can simplify it by saying that there are just three essential rules for sentence construction:

1 Every sentence must have at least one noun and one verb.
2 The emphasis should be either (a) in the middle of the sentence or (b) be divided equally between the beginning and the end.
3 The shorter and less complex the sentence, the clearer will be its sense.

In the last case, however, remember that a series of short consecutive statements makes dull and difficult reading. Interspersing shortish sentences with a reasonable number of longer ones—each consisting of a number of related clauses joined by conjunctions such as 'and', 'because', 'or' and 'unless'—can make a passage more varied and therefore more readable. However, this can be overdone too. The great pathologist Rudolf Virchow commented in the *Bulletin of the New York Academy of Medicine* in 1928 that 'The conjunction "and" commonly serves to indicate that the writer's mind still functions even when no signs of the phenomenon are noticeable'.[5]

Very lengthy sentences become so convoluted and drawn out that they are often nearly impossible to follow. Try the following example (from a letter written in 1848[6]):

> With regard to that (letter) from Dublin I can only say that your correspondent in telling you that I asserted that "pain had no effect on the mother" informed you as incorrectly as your other "*Dublin man*" who reported my opinion on the "religious objections"—on which subject you say you were induced to write your "Answer" by being informed that I was publicly advocating these so called "Religious objections" and that I had denounced you *ex cathedra* as acting in an unchristian way in advocating the abrogation of pain in labour by anaesthesia—and that the only ground you had for thinking that I did so was hearing it "very casually from a Dublin man" I really feel astonished that *you,* who must know as well as anyone, how constantly what a lecturer says is misunderstood or misrepresented, could thus admit on mere hearsay evidence a position to which you attached importance to induce you to take the trouble of writing a formal reply to arguments which never were made use of by me—I never advocated or

> countenanced either *in public* or *in private* the so called "*Religious objections*" to anaesthesia in labour, but invariably rejected that objection and many and many a time have had trouble of shewing patients the utter untenableness of such an objection—as is perfectly well known to every one here.

If you worked your way through that and still retained the message, congratulations. The probability is that you got as far as the first few lines then skipped the rest: and that is what readers will do with any of your sentences which are inordinately long.

When you do construct longer sentences be careful not to include too many qualifying or modifying words. Such additions as 'usually', 'under certain conditions', etc. may be necessary but the line between reasonable (and sometimes essential) caution and fussy legalistic hedging about is a fine one, preferably not over-stepped. To say that 'normally one may expect' a certain result is acceptable, but one qualifier is enough. By and large, taking one thing with another, other things being equal, in the normal course of events and generally speaking one can easily overdo it.

It is often said that one should prefer the active voice to the passive voice—that is, 'The nurse labelled the sample' is preferable to 'The sample was labelled by the nurse'. In that example the assertion may be true but often in science writing the passive voice may be preferable. In reporting a series of experiments the style of 'this was done prior to that being examined' is certainly less tedious (and less egotistical) than saying '*I* did this before *I* examined that'. In the same way one should generally prefer the third person to the second person when giving details of a methodology. 'This is done then that is done, followed by the other' is better (and less irritating) than '(1) Do this. (2) Do that. (3) Do the other'. The latter style always reminds me of the song 'Ol' man river', with its curt reminder of slave days—'Lift that bar, tote that bale'.

Finally, there is syntax to consider. This word is one of three (the others are punctuation and spelling) which probably frighten off more potential authors than any others, but syntax is no more than common sense in action. It merely means the way words are arranged to form phrases and sentences. If you stop to think of the logical meaning of a string of words syntax resolves itself. One should, for example, avoid the painful way of performing an experiment. 'After standing in boiling water for 30 min, examine the flask.' Or there was the title of a paper published in 1968 which appeared to herald a new way of becoming pregnant: 'Multiple Infections among Newborns resulting from Implantation with *Staphylococcus aureus*'. Think about it. In the first example it is (presumably) the flask which is to be stood in boiling water and not the operator: in the second case the newborns certainly did not 'result' from bacterial implantation, although their infections may have done. We are really back to the order of subject/verb/object/complement which we looked at earlier, but a little logical thinking will recognize such syntactical errors more easily than attempts to apply theoretical rules of grammar.

Forming paragraphs

A paragraph is a passage of writing which brings together a series of connected facts or thoughts and which divides the text into a series of more readily digested sections. The end of a paragraph marks a place in the argument at which the reader can pause and take stock, ensuring that he has grasped the subject-matter thus far before he proceeds. Generally a paragraph will begin with a 'topic sentence' which indicates what the paragraph will contain, although where an argument is in a number of related parts a new paragraph may well pick up where the last one left off, after a pause for reflection.

Because each paragraph is a separate unit capable of standing on its own, and because paragraphs may vary enormously in length, experienced fiction writers use this writing-unit to give 'pace' to their narratives; short paragraphs move the action along while long ones give a more leisurely 'feel' to the writing. The practical corollary of this for the science writer is that if he fails to divide his writing into reasonably sized paragraphs the reader will be left with a slow-moving, increasingly boring text which is (sadly) altogether too typical of much writing by scientists. Printed text without paragraph divisions is about as 'readable' as a telephone directory, yet all editors are familiar with typescripts in which a six-page article is presented as one long paragraph. Whenever a writer pauses for rest or inspiration it is almost certainly time for a new paragraph.

Punctuation

This is another absolute turn-off for the embryo author. We all think we can punctuate and we all know that nobody else does it properly. Like prowess as a car driver, an individual's punctuation is something one can rarely criticize to his face. In fact the commonest fault is not wrong punctuation so much as over-punctuation.

Punctuation merely shows which words should be taken together as units: it is a written way of expressing the pauses and intonations which distinguish normal human speech from the pale imitations produced by voice synthesizers. If in doubt one only has to read a piece of writing out loud with short pauses for commas, longer pauses for full stops, and so on. Again, however, the position can be summed up in four simple rules:

1 Full stops. Use frequently
2 Commas. Use very sparingly
3 Quotation marks. Use only for quotations
4 Everything else. Use your common sense, or avoid altogether

Full stops mark the end of sentences and, as already pointed out, the best sentences tend to be short. Commas, on the other hand, represent a lesser pause and serve to emphasize the sub-divisions of thought within a sentence. The general rule with commas is the fewer the better. If in doubt, leave them out.

Quotation marks tend to be over-used also. One should normally place single quotation marks around any passage which is quoted directly from another source and double quotation marks around any quotation from a third source which appears within the single quotation marks. Easy, isn't it?

Other forms of punctuation are generally self-evident. Colons and semi-colons are merely pauses intermediate between a comma and a full stop. Question marks come at the end of a question and exclamation marks only after an exclamation, with brackets and dashes used as an alternative to commas to mark a parenthesis—a passage inserted into the text which is capable of standing on its own—or additional information (such as comparative data) which supplements the main text. Got it? Good!

Spelling

Spelling is the third of the main worries for wary writers: and this time with good reason. Almost everyone has blind spots with spelling and there are certain words in English which positively invite confusion. Yet accurate spelling is important—especially in science—for three main reasons.

1. Incorrectly spelled words spotted by a reader distract his attention and break the train of what may well be a complex thought process.

2. There is a real risk of ambiguity. While many mis-spellings are self-evident, others are not. As an example of this the chemical L-cysteine is readily soluble while L-cystine is not. In this case an e makes the differance between success in making up a reagent and total failure.

Now, hands up all those who noticed the deliberate spelling mistake in that last paragraph. The one that did not affect the meaning but which would (if undetected) illustrate the third reason why spelling is important.*

3. Poor spelling is evidence either of semi-literacy and poor education (i.e. the writer doesn't know any better) or of careless-ness in ones writing (i.e. he/she can't be bothered to make a proper job of writing a report). In the latter case can one really trust the accuracy of the laboratory work being reported by such an indi-vidual? For example, what would you make of the examination candidate who said of a chemical that he would grind it 'with a pedestal and motor'? What he meant, of course, was 'with a pestle in a mortar': but was he being ignorant or just careless? In either case this was scarcely designed to impress the examiner (myself) who marked his paper.

In practice only some 100 words are said to account for a third of all spelling difficulties in English. A number of these fall into one of four categories and knowing the rules for them will save a lot of trouble.

1. The mnemonic many of us older ones learned at school (in the days when the 'English' taught was mainly language and only

*The correct spelling of the seventh word in the last sentence of the previous paragraph is, of course, 'difference'.

secondarily literature) rarely fails: 'i before e, except after c'. Thus one has:

BELIEVE *not* BELEIVE
but
RECEIVE *not* RECIEVE

There are a few notable exceptions to this rule, however, such as WEIGH, and even SCIENCE itself.

2. One should always drop a final e before adding to a word a suffix beginning with a vowel. Thus while one writes USE, adding -ING or -AGE converts it to

USING *not* USEING
and
USAGE *not* USEAGE

3. It is necessary to be careful of longish words which may have two sets of double letters in them. The only solution to this problem is to be aware of it and make a point of fixing such words in the mind as they are noticed. Thus one has:

ACCOMMODATE *not* ACCOMODATE
and
OCCURRENCE *not* OCCURENCE

4. Many words are written differently to the way they sound. This includes words with silent letters like:

PNEUMONIA *not* NEUMONIA
and
DIPHTHERIA *not* DIPTHERIA

It also includes less obvious examples which are made more difficult by a general carelessness in pronunciation which many people (including radio and television announcers) seem to cultivate these days rather than just fall into. We all know that the name of the city is COVENTRY and not CUVENTRY, and most of us should be able to spell SECRETARY and not SECERTARY: these are simply examples of affectation or carelessness in speech. It is less obvious that one writes

DEFINITE *not* DEFINATE
and
SEPARATE *not* SEPERATE

Once again the answer is to spot these words and remember them.

Finally on this topic here is a short list of similar pairs of words which commonly cause trouble and are best learned by heart once and for all, along with any others which you regularly trip over.

BORN = Delivered at birth	*and*	BORNE = Supported
COMPLIMENTARY = Expressing praise *or* free of charge	*and*	COMPLEMENTARY = Completing something
FOREWORD = A preface	*and*	FOREWARD = Directed to the front
PRACTICE = (Noun) Action, as opposed to theory	*and*	PRACTISE = (Verb) To carry out in action: do habitually
STATIONARY = Not moving	*and*	STATIONERY = Writing paper envelopes, etc.
METRE = A unit of measurement	*and*	METER = A measuring instrument
CONFIDANT = A person to whom you entrust secrets	*and*	CONFIDENT = Feeling or showing assurance

And in conclusion

Well-written English always sounds right. Try reading what you have written, preferably out loud, observing the pauses of your punctuation and listening to determine whether you can understand yourself. A tape-recorder may prove helpful here. Don't be afraid to alter what you have written if it is not clear, if it is potentially ambiguous or if it sounds clumsy. Use a dictionary frequently, both to check spellings and to make sure you have chosen the right word. (Earlier in this chapter I used the word 'egotistical'. Can you distinguish it from 'egoistical'.) Don't be afraid to resort to a dictionary in any case of doubt.

Achieving a pleasant and readable style is something which takes time. When reading over your writing look out for places where you have used the same word in closely situated sentences and introduce variety by changing one or the other. If you can't easily think of an alternative then browse through a thesaurus. This most valuable reference book, available in many cheap editions and to be found in all libraries, lists synonyms and antonyms—those words having the same or similar meanings and opposite meanings—and often provides inspiration for the critical writer who has run out of different words for a particular purpose.

Finally, don't expect it to be easy at first. Two hundred years ago Richard Brinsley Sheridan said that 'Easy writing's vile hard reading'[7]; and a modern corollary has been added that 'Easy reading is curst hard writing'. But it really is worth the effort.

Chapter 2
Writing essays

> Proper words in proper places make the true definition of a style.
>
> Jonathan Swift (1667–1745)
> *Letter to a young clergyman*

The first piece of science writing that most people tackle is an essay but 'mighty things from small beginnings grow' and this form of writing is the logical starting point for any study of how to write. But what is an essay? *The Shorter Oxford English Dictionary* says that it is 'a short composition on any particular subject'—which seems fair enough to most people, who have memories of their school days to guide them. But that same dictionary also calls it 'a first attempt in learning or practice' and it is always helpful to regard the writing of essays as practice for more serious and extended science writing. Of course the immediate aim of the essay-writing student or undergraduate is to satisfy the tutor who set the exercise, and subsequently to perform satisfactorily in those examinations which use this art-form. So if for no other reason, essay writing is clearly worth mastering: but there is another reason, as the dictionary indicates. If we accept that all scientists will need to be able to write clearly and effectively (and we have already established that), then a sound basis laid in student days is the surest way of working towards true literacy as a practising scientist. The rules for writing good essays are valid for writing good articles, and are best learned early in ones career.

There are three basic steps in writing an essay: planning; writing; and revising. Let us consider these in turn.

Thinking and planning

The first step in essay writing—as in all writing—is to organize ones thoughts. A title set by a tutor, or the wording of a question in an examination, should help you define the purpose of, and the breadth and depth of matter to be covered in, the composition. Having given a little thought to the boundaries, however, consider your readers. Rudyard Kipling wrote[1]:

> I keep six honest serving men
> (They taught me all I knew):
> Their names are What and Why and When
> And How and Where and Who.

And these are the companions to bear in mind when writing, for they are the questions a reader (or examiner) will ask himself. The trick of good essay construction is to forestall them by providing the answers in your essay.

So begin by sketching out an essay plan. As you think round the subject and start to assemble ideas write down on a separate sheet of paper odd words and phrases which identify the points which you must cover. The order doesn't matter at this stage—just put things down as they occur to you. When you feel you have recalled all the main points use them to draft a list of headings—and maybe sub-headings—under which you can jot down further odd words to indicate matters which will need special emphasis or explanation. Finally ask yourself a further series of questions and, as you answer them, number your headings in what seems to be the most logical order to tackle the subject; then use arrows to indicate where any diagrams, sketches, tables, etc. should go. The questions to ask are:

1 How will I introduce the subject?
2 What ideas will fit into each paragraph or section?
3 What diagrams, sketches, etc. are needed (and where should they go?).
4 Which points need most emphasis? (Underline these words.)
5 What is the most logical (and effective) sequence of the component parts?
6 How will you finish the essay?

Rule number one in all writing, from childrens' fairy stories to Nobel prize-winning theses and monographs, is that every written work requires three things—a beginning, a middle and an end. You cannot miss out any of these. Only by preparing a plan of campaign—and sticking to it—can you hope to maintain control of the monster which you may be creating. Readers—and especially examiners who may have to go through forty versions of the same thing—either lose interest or get cross if items are introduced out of a logical sequence, or if there are large sections of irrelevant material, or if some vital point has been forgotten.

So what is a logical sequence? Here I must 'jump the gun' a little and refer to the conventional order of approach in a science article. Leaving out peripheral details, the skeleton is known as the IMRAD structure: **I**ntroduction, **M**aterials and methods, **R**esults and **D**iscussion. We will return to this in the next chapter but essentially it seeks to deal with the reader's unspoken questions, **Why**? **How**? with **What** result? and **What** does it all mean? Whenever possible try to follow the same order in an essay: it is a sound model for all science writing. Remember, however, that your discussion must not be merely a repetition of what has gone before. Discuss means 'examine by argument', not 're-hash your results'.

Finally, don't forget to plan a proper ending for the piece. 'Introduction' implies a beginning; there is no comparable heading called 'The End'. Conclude your piece positively and finally and don't let it just fade away.

So now you can get down to the next stage.

Writing

The hardest part of any writing is picking up your pen. Once you have taken the plunge and started putting words down on paper it becomes easier: breaking the ice is the difficult bit and the first word is the hardest to write.

Once you have started, try to write in one continuous spell, using the words which come most readily to mind (but being careful of syntax, style and the points covered in Chapter 1) and following the order of your essay plan notes. Do be careful to stick to your remit. There is always a tendency to wander from the point—especially when the words are flowing easily—but this will either confuse your reader (and probably yourself too) or so annoy an examiner that he will simply draw a line through it and disregard it. Either result is counter-productive.

Do try to be consistent in your use of units of measurement. Generally one expects to find units of the *Système Internationale* (SI) but whatever system you use, stick to it. Don't mix ml and cm^3, or mol/l and M. Speaking of measurements, do avoid a spurious impression of accuracy: phrases like 'approximately 123 μl' are simply a contradiction in terms. Approximation does not run to three decimal places of a millilitre.

You must be careful, too, when indicating dilutions—this is something which more people get wrong than ever get right. If you mix one part of solution A and four parts of solution B you have diluted solution A one to four, which is one *in* five. Both a colon and a diagonal stroke are signs used to indicate proportion, so your dilution could be expressed as

1:4 OR 1/4 OR 1 in 5

The common mistake is to assume that the first two symbols mean 'in': they don't. To avoid any ambiguity on this point always use the unarguable form of '1 in *x*'.

Also be careful with abbreviations. As I pointed out in Chapter 1, any given abbreviation can mean one thing to one person and something quite different to another. 'AA' can mean 'Auto-Analyser', 'Anti-A', 'Automobile Association' or 'Alcoholics Anonymous'. Even a widely-used abbreviation like 'NAD' has at least 72 grammatically possible scientific interpretations. If you must use abbreviations (and their use is often simply verbal laziness) you must spell out the words in full at the first usage, followed by the abbreviation in brackets: you may only then use the abbreviation on its own afterwards.

Finally, try to be tidy and legible. Essays are nearly always handwritten and if an examiner can't read it he can't give credit for it. It is not that difficult to write legibly and it is an illuminating—and chastening—experience to hand a piece of your own handwriting around half a dozen of your fellow-students with a request for frank criticism of its legibility. Copper-plate or italic script is not looked for by examiners but simple readability is. Exaggerated flourishes, steep slanting of letters and tiny cramped words are as bad as simple scrawl. A few extra minutes taken to write a piece neatly are worth more than any amount of time saved by rapid scribbling which no-one else can read. It has even been known for students to be unable to read their own writing a day or two later. This must be the ultimate in non-communication.

Revising and reviewing

So having written your essay what do you do next? Well there are essentially two processes at work in written communication:

1 **You** express the thoughts in your mind by selecting appropriate words.

2 **Your reader** converts your words into thoughts in his mind.

The difficulty is in ensuring that your original thoughts and those which end up in your reader's mind agree. If this is to happen then not only must you select the right words but it is essential that you carefully check over all that you have written to make sure that nothing has gone wrong along the way and that the finished piece of writing actually conveys what you meant it to. It is not difficult when writing—especially if the words are coming faster than the pen can cope with—to write down words which either fail to convey your meaning or which convey the wrong meaning; generally through ambiguity, faulty syntax or misleading (or missing) punctuation.

To overcome this it is best if you can put your essay aside for a day—or a week—or two and then check it over with a fresh eye when faults will jump out to meet you which would probably pass by un-noticed if you simply read over the piece immediately after you finish writing it. If you can't put the piece away for long (as in an examination), at least clear your mind by setting it aside while tackling something else (like another question) and then check it when the thoughts you had when writing it are no longer ringing in your mind.

The process of checking should be twofold. First, a rapid read-through to see if the sense of the piece comes over: this will bring to light any glaring omissions or mis-statements. Second, a more careful reading with a conscious search for ambiguities, etc. Beware of having missed out words due to your mind having been racing ahead of your pen. 'The results were significant' is only three letters different from 'The results were *not* significant', but the meaning is exactly opposite.

Revising any piece of written work is not an optional extra when time permits but an essential and integral part of the writing process. Probably more examination candidates fail as a result of

Stage 1

Activation of the coagulation system, with production of and subsequent lysis of fibrin, is a feature of many physiological and pathological conditions. The measurement of circulating fibrinogen/ fibrin degradation products which represent the end point of lysis, has proved to be an effective means for monotoring this process. One of the main advantages of the technique described in chapter 5, is that it strikes an ideal balance in that it can be used for the "one off" emergency fibrinogen or F.D.P. assay or multiple, large batch assays for monitoring of such disease processes in a number of pathological states.

Stage 2

Activation of the coagulation system, with production ~~of~~ and subsequent lysis of fibrin, is a feature of many physiological and pathological conditions. The measurement of circulating fibrinogen/ fibrin degradation products (which represent the end point of lysis~~,~~) is ~~has proved to be~~ an effective means for mon~~o~~itoring this process. One ~~of the main~~ advantage~~s~~ of the technique described ~~in chapter 5~~ is that it ~~strikes an ideal balance in that it~~ can be used for single ~~the "one off" emergency fibrinogen or F.D.P. assay~~ or multiple~~,~~ ~~large batch~~ assays for monitoring ~~of such~~ disease processes in a number of conditions ~~pathological states.~~

Stage 3

Activation of the coagulation system, with production and subsequent lysis of fibrin, is a feature of many physiological and pathological conditions. The measurement of circulating fibrinogen/ fibrin degradation products (which represent the end point of lysis) is an effective means for monitoring this process. One advantage of the technique described is that it can be used for single or multiple assays for monitoring disease processes in a number of conditions.

Fig. 1. Shortening a piece of writing by editing.

lack of careful revision of their written work than do so due to inadequate preparation.

The essential concept is that of self criticism. This can best be learnt by practising criticism of others' writing. You can learn to recognize superfluous words, bad grammar, misleading syntax and faulty punctuation by photocopying articles in newspapers, magazines and medical (rather than science) journals and then, with a red pen, trying to improve or shorten them: then apply the same principles to pieces of your own work. Figure 1 (taken from a research thesis) show how a piece of writing can easily be cut by a third without affecting the meaning. Writing at unnecessary length (especially when under a time constraint, as in an examination) has two effects: it wastes your time and it confuses the reader. Most people don't realize they do it, but removal of the chaff leaves a better quality wheat. Revision of written work is really no more than quality control.

In the long run there is only one way to improve your essay writing:

PRACTICE—PRACTICE—and more PRACTICE

Chapter 3
Writing science papers for publication
1: The body of the paper

'Where shall I begin, please your Majesty?' he asked. 'Begin at the beginning', the King said, gravely, 'and go on till you come to the end: then stop'.

Lewis Carroll (1832–1898)
Alice in Wonderland

We live today in a world of shortages; energy, foodstuffs, raw materials, money—the list seems to be never-ending. Of one thing there is no shortage, however: on the contrary, there is a positive glut of scientific information. *Index Medicus,* the major indexing periodical for the biomedical sciences, currently lists 26 000 journals in this subject area alone, with some quarter of a million individual citations (i.e. scientific papers) each year. Faced with such a high level of production, journal editors have to be very selective indeed and most reputable journals reject more than half of the items submitted to them: in some sciences (e.g. physics) this might rise to over 90%.

The reasons for acceptance or rejection of a paper are many. You will, of course, improve your chances considerably by carefully reading the 'Instructions to Authors' of your target journal—and following them. In general, papers are just as likely to be rejected for poor presentation as for poor scientific content. Remember, authors of scientific papers are in a highly competitive market place in which only the very best will survive. The corollary of this is that to be successful a paper must be well-written and well-presented, as well as reporting sound scientific work. The work itself is outside the scope of this book, but let us look at the preparation of the paper.

Format

All science papers have a standard format. One may rejoice in this or deplore it, but it represents the result of three centuries of developing tradition, editorial practice and publishing procedures. It is the system, it works, so don't try to 'buck' it.

The basis of the format—known by the acronym IMRAD—is an attempt to answer in advance (and in logical order) the questions which a critical reader would wish to ask. 'Why did you start, what did you do, what answer did you get and what does it mean anyway? That is a logical order for a scientific paper.'[1] These questions are answered by producing sections in each paper devoted to an Introduction, Materials and methods, Results and Discussion (hence IMRAD). This is only the core of the paper,

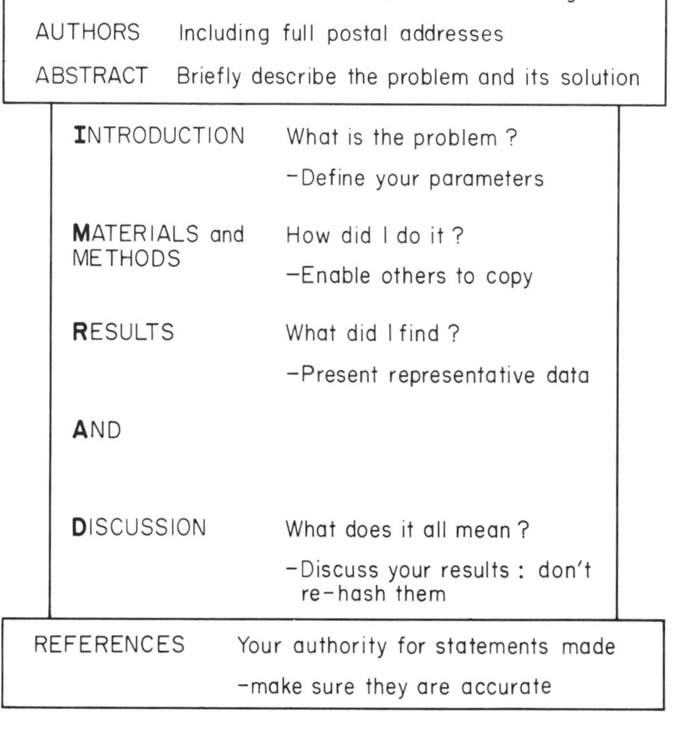

Fig. 2. The IMRAD sandwich format for science papers.

however, and there are a number of peripheral items which sandwich it and provide additional information and evidence. The total format is shown in Fig. 2.

Planning

As with essay writing, an outline plan of action is the natural starting point. Once you have decided to write a paper you will begin to collect together your data. Do this in an orderly way using record cards, or folders, for each section of the paper. Tables, graphs, diagrams etc., can be sketched out at this stage and this will help you to see if there are any gaps in your work or any conflict between different sets of results. Shuffling the contents of the files will help you get an overall picture of how the finished paper will look and this, together with the convenience of having all of your material to hand and in order, will mean that you can sit down and write in one continuous spell (as with essay writing), so maintaining continuity of style as well as of thought.

It is during this preliminary planning stage that you should look critically at the arguments which you are advancing and at the evidence you will produce in their support. Are there any flaws or inconsistencies? Does your case conflict with other published work? If so, can you defend it—or must you, in all honesty, admit that your case does not stand critical examination? (If this is the case then you must ask yourself whether you should be writing the paper at all.)

Also at this stage consider the presentation of your data. Are tables more effective than graphs, or vice-versa? Which data are essential and which can be left out so that they do not confuse the issue? Will your tables and graphs be laid out with the correct columns, scales and axes to make your point most effectively? Where is the most logical—and effective—place for each item to be referred to, and to appear? Should tables and/or graphs be grouped together for maximal effect?

Finally—and most importantly—make sure that you have a complete and accurate set of index cards for your reference citations. You should have made these up while performing your work, as you referred to each previous publication for information or ideas. If you did not do so then do it now, before you start writing. And make sure that all your reference data are accurate and written in the right order for the system used by your target journal (see Chapter 4).

You should now be ready to begin writing, so before we look at each item of the paper in detail let us consider the mechanics of the task.

Sitting down to write

Before you start writing you should give some thought to how you will actually put pen to paper. Apart from a pen you will find a pencil handy as well as a ruler and a bottle of white ink (correcting fluid). You will find it easiest if you write on A4 size paper (297 mm × 210 mm) of the type called 'narrow feint'. This paper has lines approximately 6 mm apart: write on alternate lines and you will then have ample room to make corrections, additions, etc. You can use your pencil to draw a 25 mm margin at the left and to indicate in this any instructions for your typist—or reminders for yourself of points to be chased up.

It is obviously best if you can sit down at a desk (or kitchen table) where you have plenty of space to spread out your writing materials, laboratory notes, reference books, etc. Do not make a fetish of this, however. Be prepared to write wherever (and whenever) you are free of other distractions and are in the mood. Much of this book was written on trains and aeroplanes, or with a television set blasting away across my living room. The main thing is to be comfortable and to have your mind concentrated upon the topic of your paper. Try to write in long continuous spells but don't go on once you have become physically or mentally tired.

So now let us look at the different components of your paper.

The title

This is more important than most people realize and usually far too little attention is paid to it. In general, titles tend to be indicative—that is, they refer to the subject matter rather than to the conclusions. Occasionally, however, titles may be informative, summarizing the paper's main conclusions in a few words. Informative titles are difficult to write effectively (try one if you don't believe me), and they are usually found in the popular press rather than in serious academic journals. While you may come across it in a newspaper—or even a popular science or medicine review magazine—few science editors would accept a title on the lines of 'A New Cure for Cancer Described'. It is too dogmatic and begs too many questions.

While the title of a paper will be seen by tens of thousands of people only a small proportion of these will read the paper itself: so if you wish to attract a wider audience you must choose every word with care. The aim is to use the fewest possible words which will adequately indicate what follows. Most science articles become known to a wider audience by means of the indexing and abstracting services—many highly reputable journals have a circulation of only two or three thousand copies per issue—and in any index it is the first few words (no more than six) which have an impact on the reader. Because there are so few words available in prime positions, avoid wasting them on definite and indefinite articles and irrelevant words. 'The Development, Evaluation and Application of a Sensitive Radioimmunoassay for Human Serum Rhubarb' gains nothing from the first eight words, which are surplus.

Over-long titles have been around a long time and they often resemble an abstract of the paper itself, as with an article published in the *Journal of the Royal Microscopical Society* in 1896 'On the addition to the method of microscopic research by a new way of producing colour-contrast between an object and its background or between definite parts of the object itself'. How would you index that one? Things have not improved a lot in some journals, as illustrated by the title of a paper appearing as recently as 1980 on 'Therapeutic effectiveness and safety of outdated human red blood cells rejuvenated to restore oxygen transport function to normal, frozen for 3 to 4 years at −80°C, washed, and stored at 4°C for 24 hours prior to rapid infusion'. One is reminded of the two students in a university library. One asked the other if he had read the professor's latest paper, 'No' was the reply, 'I haven't even had time to read the title yet.'

One can overdo things in the opposite direction. A paper entitled 'Studies on enzymes' is so vague as to be meaningless. What kind of studies on which enzymes, with what intention? The exception to this criticism is when the subject is a closely-defined one and the paper is a review article covering the whole subject. One such paper had a one-word title—'Metachromasia'.

A final word on titles: do avoid trade names and, above all, abbreviations. Think of the indexer again. Where would you put (or look for) NaCl? Under 'NaCl' or 'Sodium chloride'? Abbreviations which are not internationally standardized can in any case mean all things to all men.

Authors

Most editors have a healthy suspicion of multi-author papers when more than two or three names are listed. It has been well said that 'two can write an article together, possibly three can, but no more. Six people can no more write an article than six people could drive a car'.[2] Of course, the word 'author' need not mean only those who put pen to paper: it can also include those who took a positive role in the planning, execution or analysis of the results of an investigation. However, the question may then be asked whether such assistance is not better referred to in the 'Acknowledgements' than by a spurious 'authorship'.

Two practices which are indefensible—but nevertheless widespread—are inclusion of a name amongst the 'authors' solely because the individual is either head of the department in which the work was performed, or because he or she is an 'up-and-coming' young research worker for whom extra 'publications' on his or her curriculum vitae have a practical value in enhancing promotion prospects. *Medical Laboratory Sciences* is one journal which requires a signed statement from each 'author' of a submitted paper that he or she has 'contributed directly to the planning, execution or analysis of the work reported or to the writing of the paper'.

The ultimate in multiple-authorship has probably not yet been reached. I know of one medical paper with 23 named authors (plus 'other investigators') but it is said that papers with over 90 authors' names have been seen in physics journals. Don't risk being laughed at by getting into that league. If you must name two or three authors do make sure, before you even write the paper, that you are agreed upon the order in which you will appear. Alphabetical, seniority, or proportional to the work done? Personally I prefer the latter; but don't leave it to argue about afterwards. You stand to lose more friends that way than many another.

Addresses

Because readers frequently wish to write to authors (usually to request offprints of their papers), and also because it may be relevant to the credence given to the work reported, it is necessary to supply a departmental postal address and that given at the head of the paper is normally of the institution in which the work was performed. When an author is no longer at that address it is usual to give his or her present address as a footnote to the front page. In the case of work reported jointly by authors from two or more institutions each address is given, in the same order as the authors are listed, each author and his address being linked by asterisks, 'daggers' etc. In every case it is necessary that the address is the full version used by the postal authorities and includes the postal (or Zip) code. Please don't forget to add the country. I live in Aberdeen, Scotland, but I know of at least five other Aberdeens around the world. Birmingham could be in England or in Alabama USA. Even London may be the one in Ontario, Canada. The lesson here is never to forget (what may seem to you to be) the obvious.

The 'abstract'

The Abstract is what many people (not always correctly) think of as a Summary and is intended to allow readers to identify quickly what the paper is about. This part of the paper is one of the more difficult to write. Anyone can get a message across in five pages of writing: it is very much more difficult to do it in five lines.

There are two views of how an abstract should be written. One says that it should just contain a précis of the introduction and the author's conclusions, while the other says that it should be a sort of mini-paper, outlining methods and results as well. You will have to choose for yourself which type of abstract you want—perhaps guided by the practice of your target journal. Be warned, however, that the second type is by far the harder to write within a given number of words. Try not to exceed 100 words: if you are much in excess of this then put it away for several days before trying to edit it into a more acceptable length.

The Abstract must not contain any information or conclusions which are not in the rest of the paper, nor should it contain any unidentified abbreviations, reference citations or trade names. If you find yourself in trouble writing your Abstract then remember the example of a very famous scientist who summarized his work simply as '$e = mc^2$'.

The introduction

Dylan Thomas started his classic *Under Milk Wood* with the words 'To begin at the beginning': and that is the point we have now reached. But an 'Introduction' is more than just a beginning: it must set the scene for what follows. This is the author's opportunity to provide background information and to explain what the investigation was all about: which is necessary so that readers can subsequently make up their own minds about the results and conclusions. It is the place to define the problem clearly and succinctly before going on to describe its solution.

Materials and methods

The purpose of this section lies in the well-established tradition that to be acceptable, scientific research work must be repeatable. In specifying the materials you used and giving precise details of the methods employed you are giving other workers the opportunity to do just that. The names of manufacturers of certain reagents will be important here, as may be batch numbers: the amount of detail will depend on the consistency of the individual materials. It is not unknown for one manufacturer's product to produce certain results solely due to impurities in it which are not found in other manufacturers' (supposedly) equivalent items. Equipment, too, may vary from maker to maker. However, it is not sufficient to cite only a name: what is familiar in England may be unheard of in Australia or Romania, so full postal addresses are also important. Don't underestimate the importance of these details: referees often test methods in their own laboratories and frequently find that they don't 'travel' well and can not be repeated. One author once suggested that I visit his laboratory in Hong Kong to see how a staining method worked using his local tap water. Citing the pH may have been a much simpler—and cheaper—solution. A word of warning, however. Unless it is central to your work don't use trade names except in the form of a reference. If you use 5-amino-acridine hydrochloride then say so, rather than calling it 'Acriflavine'. If you used a 'Whirlimixer' then remember this is merely one manufacturer's version of a vortex mixer: and please don't coin yet another verbal-noun and say you 'Whirlimixed' a solution when you vortex-mixed it.

Do be precise about incubation times and temperatures, and be consistent with your units of measurement: if you start referring to a 15 g/l solution don't go on to speak of it elsewhere as 1.5%. One common fault to be strenuously avoided is to claim to have sterilized something at x lbs/in^2. It is temperature which kills bacteria, not pressure, and if you appear to be unaware of this your general credibility is called in question.

A word about style. If you are giving details of something like a staining technique it is too simple to list the instructions in the second person—'add this; leave for 3 min; wash off; add the other'. There is little literature less literary (and more dull) than this style of writing. It may be harder and take longer to convert such details into prose but this will result in a much more readable passage which will make more impact on your professional peers.

Finally, don't be tempted to mix your results in with this section. They deserve a separate section of their own, so let us move on to it.

Results

This is the really significant part of the paper, where you provide the evidence upon which your research stands or falls. However, your results must come across clearly. This means that when you have a mass of data you will need to be selective and produce sufficient representative information to substantiate your argument, but not so much as will confuse the issue. In most areas involving numerical data the best form of presentation is to use either graphs or histograms, but don't make the mistake of duplication. Give your results in the text, or as tables, or as graphs, or as histograms, but never give the same results in more than one form. Above all, avoid redundant text which merely repeats the information in your tables or illustrations but at much greater length.

Now a word about statistics. Very few scientists are more than very amateur statisticians. A brief undergraduate course too often leaves laboratory workers thinking that a knowledge of the meaning of Chi squared and standard deviation makes them instant statisticians. Nothing could be further from the truth. Indeed, statisticians are notorious for their own frequent disagreements between one another: this is not an arena for amateurs to enter unprepared. Nor can one get away with the light-hearted quasi-statistical approach. There is an apocryphal story of the scientist who reported a trial of a new drug claiming that one-third of the mice used in the experiment were cured by the drug, 33.3% were unaffected by it—and the third mouse got away. He had obviously never heard of statistically significant population sizes. The only possible course of action if your data require statistical analysis is to consult a statistician: you can be reasonably sure that the editor to whom you submit your paper will do so.

The 'Results' section of your paper, although important, does not need to be long or complex. In some papers it is possible to say 'The results obtained are shown in Table 1' and that is the section complete: usually at least some text is called for, but it need not be extensive. What does require more extensive attention, however, is your discussion.

The discussion

This is probably *the* most difficult section of any paper. 'Discussion' is defined as 'to examine by argument' and properly it should consist of an examination of your case supported by your results: it must not be mainly a re-hash of these, as is all too often the case. Discuss your results, don't repeat them.

In this section of the paper it is necessary to highlight which of your results are most significant—and this means both statistically and in the dictionary sense of 'having meaning and importance'. One should also point out any deficiencies or inconsistencies in your work. This is simple self-honesty, for if there is any inconsistency of which you are aware you can be certain that your readers (and critics) will not fail to see it also, and rush to point it out. You should also compare your results with those of any comparable previous investigations, and then discuss their theoretical considerations.

Many people seek to include a section of 'Conclusions' at this point but this is not general practice. Your 'Discussion' should itself be structured so that it leads naturally to a point where you can draw conclusions from it. A 'natural order' for a discussion might be as follows:

1 Outline previous published results
2 Compare your present results with these
3 Highlight any significant new results
4 Point out deficiencies and/or inconsistencies, either in your present results or in previous published results
5 State your hypothesis. Make it clear what you are trying to prove—or disprove
6 Support this by your new results
7 Draw conclusions from your argument
8 Conclude by summarizing your argument or your recommendations in a single—preferably short—sentence.

You will notice that in this 'battle plan' I have referred both to new results and to what it is you are trying to prove. Perhaps the necessity for both of these elements is why so many 'Discussion' sections are poorly written and lead to the rejection of the paper by editors who, amongst other things, have a filtering function to perform. A scientific paper has been described as 'a written and published report describing original research results'.[3] Many workers try to produce a paper around trivial investigations or routine tests which do not represent true research—i.e. 'study and investigation to discover new facts'. If there are no new facts then there has not been any successful research. Most science journals contain sections devoted to short (mainly technical) communications and to review articles, and these fulfill very valuable functions—but they are not original scientific papers. It is in writing your 'Discussion' that the true nature of the paper comes clearly to light and your work is exposed for what it truly is: either research *or* development *or* review. The latter two types of paper do not need a formal 'Discussion', although they may well follow the same general format as an original article reporting research findings.

Remember the advice given in Chapter 3 that every piece of writing requires a beginning, a middle and an end. The 'Discussion' is the effective end of your paper so pick your words and phrases carefully. You want your work to go off with a bang, not fade away with a whimper. Aim to finish positively, decisively and forcefully. And then stop writing.

Chapter 4
Writing science papers for publication 2: The 'peripherals'

> Three hours a day will produce as much as a man ought to write.
>
> Anthony Trollope (1815–1882)
> *Autobiography*

Of course, you can't really stop writing at the point which we reached at the end of the last chapter. Nor have you really reached 'the end' of your paper, for although the body of it may be complete there remain the rest of the 'peripherals'.

The acknowledgements

These represent a small but important opportunity for the author(s) to record the help of others who have contributed towards the success of the investigation.

Many of those whose names appear as co-authors might more reasonably receive acknowledgement in this place instead. I recall a paper some years ago with a 'co-author' known to me whose sole contribution to the work had been to drive into the country in order to collect samples of blood from members of a family which was being investigated. This was surely a case for an 'acknowledgement' at most. There may indeed be very many people who have helped your work—colleagues, your superiors, assistants, whoever typed your manuscript, etc. In particular, this is the place to refer to any grant-awarding body to which you are beholden or to any commercial enterprise which has helped you with reagents or equipment.

There are two general rules to follow when making acknowledgements to individuals. The first is not to trivialize your paper by being too 'matey' in your comments. They may well be your friends but something like 'I'm grateful to old Bill Riley for giving me some samples and to Mary McKay for holding the rabbits' is just too informal. Secondly—and more importantly—do make sure you have the prior permission of anyone you name in your 'Acknowledgement'. After all, a colleague may be chary of being made a responsible party to your work if he has not seen it: and he may not agree with your conclusions even if he did help you out with information or materials.

References

And so we come to another very difficult part of any paper, which is almost always badly done. It is essential to the academic credibility of your work that no statements are made or the results of other work referred to unless these can be substantiated: hence the system of reference citations. There is a trap here for the unwary, however. If you make an assertion that Dr John Doe obtained 100% of positive reactions when testing 512 Lithuanians in 1943 be sure that you have personally seen Dr Doe's paper and confirmed the facts: it is not enough that the standard textbook by Professor Superman, published in 1980, quotes these figures, nor that they are to be found in *The Dictionary of Kidology*. Academic workers distinguish clearly between three levels of source material:

1 Primary sources. The original reports of original research, written by the research workers themselves.

2 Secondary sources. The information from primary sources as reported in textbooks—and sometimes in review articles—by other workers who (presumably) have themselves read the originals.

3 Tertiary sources. Collected information from secondary sources, as found in encyclopaedias, dictionaries, etc.

When citing references to other work it is safe to use only primary source material: you are then responsible yourself for accurate reporting. Ideally, never cite a reference to any published work which you have not read yourself. Sadly, it is not uncommon for particulars—not only of the work but even more of the bibliographical data—to become altered on transmission through secondary (and even more through tertiary) sources. I have more than once tried to check a reference to an article alleged to have appeared in (e.g.) volume 38, 1981 of a journal only to find that volume 38 appeared in 1979 and 1981 was volume 40: and the article was to be found in none of these volumes or years. References which don't refer are worse than useless: they call in question the accuracy and reliability of the whole of your work. The only way to guard your credibility is to check, double check and then check again every detail of every reference you cite. Of course you can make things easier for yourself (and your reader) by taking care only to refer to work which is significant and relevant. Don't cite unnecessary references merely for the sake of it.

The format of each reference may vary depending upon the style adopted by your target journal. If in doubt, however, it is best always to give all the information possible. Editors are more prepared to modify your text by deleting words and figures than by inserting them. As a general rule you should give the author's surname and initials; the full printed title of the paper (or book); the full title of the journal containing it (NB don't abbreviate the journal title unless you are certain of the correct system of abbreviation) OR the place of publication and name of the publisher of a book; the year of publication; the volume number (of a journal); and the first and last page number of the article or section cited. When you are citing work originally published in a foreign language (which presumably you are able to read) use that original language for the reference citation, although in the case of non-Roman script (e.g. Arabic, Japanese) a translation is acceptable if followed by '(In Japanese)', etc. If you are using someone else's published translation say so, and give the reference to it as well as to the original. The essence of the system is to avoid guesswork.

It will be necessary, once your paper is complete, to check that there is full agreement between the references cited in the text of the body of your paper and the list given at the end. The actual method of citing the references is another minefield, however, over which bitter battles are still being fought. There are three basic systems in use and the proponents and opponents of each give no quarter in the fight. Use whichever system is appropriate for your target journal (see Chapter 6) and never mix systems. The details of the principal systems are as follows.

1: The 'Harvard' system (names in the text)

This is the oldest of the current systems and is now probably the least used in medical and science journals.

In the text, when a reference is to be given, the surname of the first author is given followed by the year of publication: when there is more than one author the first, second or third name (depending upon editorial policy) is followed by '*et al.*' The whole reference is placed inside brackets. In the list of references the entries are then placed in alphabetical order of the first named authors.

The advantages of this system are that it is simple for the author when writing his paper, and readers can instantly recognize the identity of any eminent author whose work is cited. Disadvantages are that readers find the text so broken up by the length of the inserted references that it can become very disjointed and difficult to read. If an eminent author is cited whose name is not the first (second or third) listed in a paper it will not be immediately obvious and in any case many first-class papers cannot be recognized by this rather dubious method.

2: *The alphabetical numerical system*

This system is most popular amongst North American journals, in some medical journals and in periodicals in the arts and social sciences. It is the system used in British Standard 1629:1950, *Bibliographical References*. The list of references is placed in alphabetical order of the first named authors and this list is then numbered from 1 upwards: the numbers are then entered in the text at appropriate places.

Advantages are that the author can update his list of references without disturbing the *position* in the text of all the rest of the numbers, although the numbers themselves will need to be changed, and that readers can readily see from the list whether the work of a particular author has been cited (but only if he is the first—or sometimes the second or third—named author). Disadvantages are that the reader cannot readily trace a listed reference in the text by its number and that the author cannot readily check through the typescript to ensure that all the reference numbers have been included and are in the right place: hence there is an increased risk of error, especially in papers with a large number of references.

3: The sequential numerical system

This is the most popular system with most European and North American biomedical science journals.

When the first reference is cited in the text it is given the number 1, the next reference is numbered 2, and so on. The list of references is then given numerically, without regard to alphabetical order.

The advantages of this system are that it is very easy for the author, who can write his list of references as he writes the paper and insert them in the text as he proceeds. It is easy to check that all the references have been noted in the text and the reader can follow the references in order while reading the paper. Disadvantages are that making any addition to the list of references upsets the whole numbering system and that the reader cannot so readily identify the presence of a specific author amongst those cited.

One helpful variation of the sequential numbering system which is used by some journals[1] but is not yet widespread, is the ability to include other information such as the names and addresses of manufacturers and suppliers of reagents, equipment, etc. In the past, normal practice—regardless of whichever reference system is used—has been to include such information either in parentheses, in the same way that names and dates are cited in the 'Harvard' system (with which it shares the same disadvantage of breaking up the text), or as footnotes on the relevant pages (which are expensive to set and create problems when the author wishes to refer more than once to the same information). With sequential numbering one can insert these details as they arise and refer back to them simply by re-quoting the appropriate number(s). At least one journal (*Medical Laboratory Sciences*) has used this variation on the system with success for some years.

Style of reference citation

The way references are laid out—use of capital or lower-case letters, punctuation, the order of the component parts, what is included or left out—varies enormously from journal to journal, system to system, and with the passage of time. In the first paper I ever published, some 30 years ago, one of the references appeared thus:

MALONE, R.H., & STAPLETON, R.E. (1951). *Brit. med. J.* 416.

Today, in the same journal, it would appear as:

Malone RH, Stapleton RR. Blood grouping on blotting paper. *Brit. med. J.* 1951; **i**: 416.

Because styles vary so much the only safe course is to read carefully the 'Instructions to Authors' of your target journal, observe the exact style used in papers in the current issue, and follow that. Of course, if you are submitting your paper to a biomedical science journal which adheres to the uniform 'Vancouver style' for presentation of typescripts (see Chapter 6) then you should follow the reference citation system of the US National Library of Medicine, as used in *Index Medicus*. The second example quoted above is written in that style.

Unpublished data

A rather different situation occurs when you wish to refer to hitherto unpublished work. This may take the form of unpublished reports which were prepared for the internal use of an organization; work which has been submitted for publication elsewhere and which is still under consideration; work which has been accepted for publication elsewhere but has not yet appeared; and information given to you (verbally or in writing) by another worker.

In the case of private reports it will of course be necessary to obtain permission from the organization concerned before you may publish its data. It would then be referred to in the text as (e.g.) 'Unpublished report by sub-committee of Council, to Institute of Lobster Farmers (1981)'.

Work which is under consideration for publication cannot be referred to as 'Submitted to *Journal of Lost Causes*': the editor of that journal may recoil with horror from the paper so the reference never will refer to it. Such items should be cited as (e.g.) 'Orwell, G. Unpublished observations: 1984' By contrast, work which has been accepted should be referred to in the normal way for published papers only with the year, volume and page numbers of the journal replaced by '(In press)'. (Very often, by the time you come to correct proofs of your paper it will be possible to enter the full details.)

Information given to you personally is cited as (e.g.) 'Chancer, A., Personal communication'. However, journal editors have different ways of dealing with these references: some list them with the other references while others merely place them in parentheses in the text. Some editors may also wish to include the address of any person so cited. In any case you will be expected to have the written consent of the individual cited for his or her name to be used. As with 'Acknowledgements', your colleagues may not necessarily wish to be connected with your work.

Illustrative material

Almost certainly you will wish to illustrate your paper by including numerical data—either in tabular form or as graphs or histograms—and quite possibly use diagrams to illustrate the construction, plumbing or wiring of a piece of apparatus. You may also wish to include photographs of microscopic preparations, etc. It is unfortunate that so many authors look upon this aspect of their work as something which any dilettante can expect to tackle successfully as a sideline to a career as a scientist. The graphic arts are a professional specialism in their own right and the best way of tackling the preparation of such material is to consult a professional illustrator. If you have the good fortune to work in a university, college or hospital with a graphic arts department you need do no more than familiarize yourself with the specific requirements of your target journal and then contact your friendly neighbourhood graphics man. If you are less fortunate you will need to do the best you can between yourself, your typist and a camera. In this case you will need to restrict your illustrative material to straightforward tables and graphs—unless, that is, you or a colleague (or friend, or relative) has the necessary talent to produce photographs or drawings to the required standard. As a guide to what those standards are we shall look briefly at each form of illustrative material in turn. First, however, let us consider what types of illustrations are (and are not) called for.

Numerical data are best presented as tables when you need to present specific results for comparison with each other. On the other hand if you wish to illustrate trends, or ranges of figures, a graph is more informative. For describing the details of a piece of apparatus or for showing electrical or plumbing circuits a diagrammatical or stylized drawing is the clearest (Fig. 3). Microscopical preparations and (occasionally) the general layout and appearance of apparatus call for photographs—although these must be of very high quality for reproduction and reduction. (NB, Cartoons are always out of place in serious academic writing.) Apart from tables, which are produced on a typewriter, all other illustrations fall into one of two categories—line or half-tone.

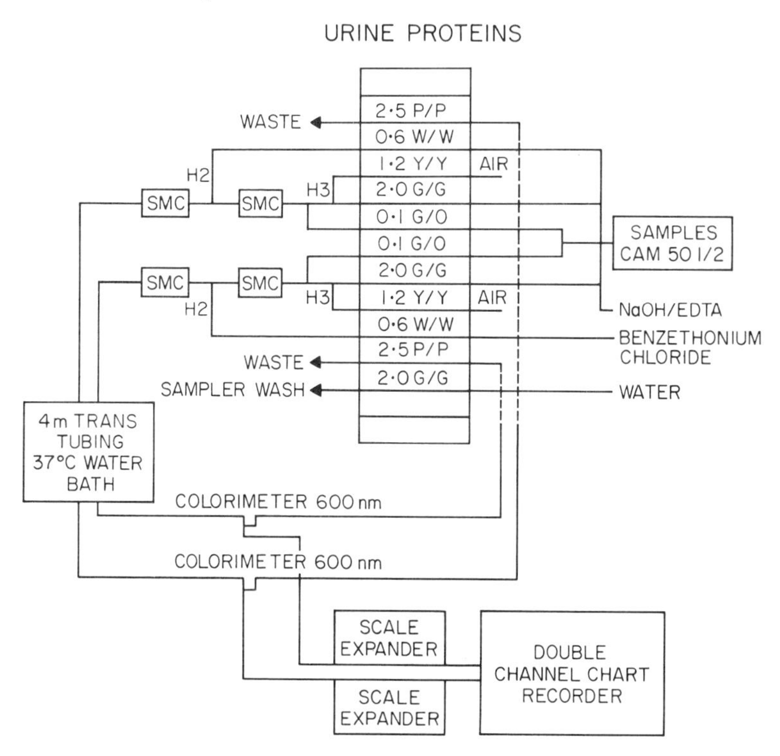

Fig. 3. Plumbing diagram for continuous-flow automatic analyser (line diagram).

Line-copy consists of black and white artwork: a graph is a typical example. Line drawings are best made on thin white card with India ink and should be made about two or three times larger than they would appear in your target journal. Tracing paper may also be used. Remember that not only the overall dimensions of the drawing will be reduced in print but also the thickness of the lines themselves. This means that you should make the lines relatively thick if they are not to appear like the threads of a spider's web on the printed page. You should not include any lettering on a line drawing but indicate this either on a photocopy or a translucent overlay; the printer will set the lettering in type or the publisher may have the artwork lettered for you. In graphs, do not include the cross-lines but just the curve and the axes (Fig. 4), maybe boxing in the whole graph with a plain frame. For reasons of safety in handling and in the mail do not produce drawings on paper or card large than A4 size (210 mm×296 mm) unless the artwork can be rolled and transported in a cardboard tube. If you really can't produce a suitable original diagram some publishers will have one drawn from a 'rough': do not presume upon this, however, but consult the editor of your target journal first.

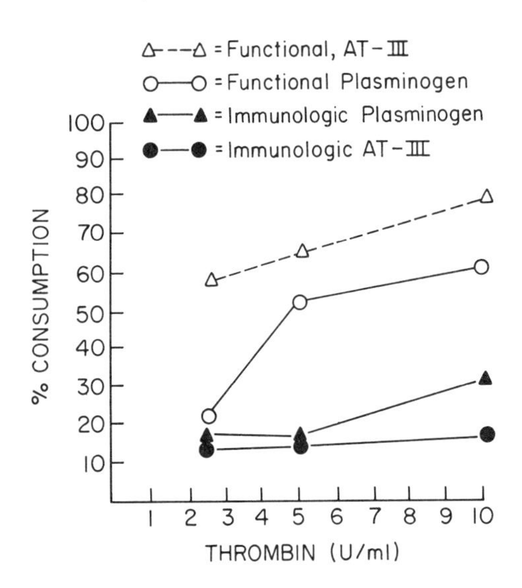

Fig. 4. A simple graph (line diagram).

Shading and stippling are useful for illustrating material where shades of grey are needed—for example histograms and drawings needing contrast (Fig. 5) or those showing animals or anatomical subjects. Nowadays artwork of this type is no harder to produce than ordinary line copy provided the shading is not too fine.

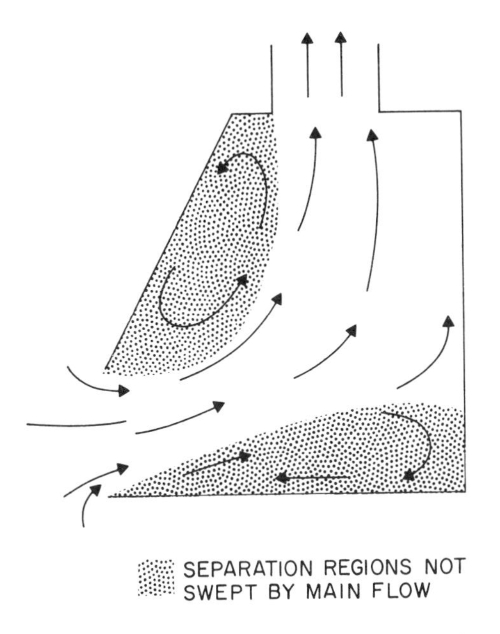

Fig. 5. Vertical cross-section of a safety cabinet (line diagram with stippling).

'Shading' is applied to the basic line drawing by means of pre-printed line tones consisting of a slightly adhesive printed film on a transparent backing sheet (e.g. 'Letraset'). A piece of the film is cut to the approximate shape required, peeled from its backing paper and rubbed down with a soft pencil or similar object. For safety, the film is best 'fixed' with an appropriate aerosol spray sold for the purpose. A range of shades is available and very professional effects can be achieved with very little practice.

Half-tone copy really means photographs: and these, for all practical purposes, will be black and white. Very few journals indeed can stand the staggering expense of printing high-quality colour illustrations: and if it is not of high quality it is not worth doing. When an editor is prepared to consider colour photographs you will need to negotiate directly as to whether he will accept transparencies (which are preferred) or prints: you will also probably find that you will be expected to bear all (or most) of the extra cost involved, which can be considerable. Few subjects cannot be illustrated satisfactorily in black and white—perhaps supplemented by textual references to the colour of (e.g.) various stained cells or tissue structures.

The basic rules for all photographs for reproduction are that they be in sharp focus, of good contrast and submitted as glossy prints about 1½ times the size they will finally appear. Many journals specify a particular maximum size of print. As with line drawings, lettering etc. should be indicated on a photocopy or a translucent overlay (which should include corner markers). Photographic prints should be identified on the back, either *very* lightly in pencil or by means of a self-adhesive label, with the author's name, the number of the illustration and the word 'Top'. *Never* type or write directly on the back of a print with a ball-point pen or a hard pencil: the imprint will show through and ruin the print for the purpose of reproduction.

Tables. A table is defined as 'an arrangement of numbers, words or items of any kind in a definite and compact form so as to exhibit some set of facts . . .for the convenience of study, reference or calculation'; and they are not as easy to produce as some people think. The basic rules are to avoid over-crowding, vertical lines and confusing column-headings. Do look carefully at the shape of your tables: a layout which is long and thin—in either direction—is unlikely to fit happily into a printed page. Tables should be clear and carefully organized so that your readers can glean the essential information from them without the need of an interpreter. In particular, each table should be complete in itself so that the legend, footnotes (if any) and the body of the table give all the relevant information without the reader having to refer to the text—which

Table 1. Example of improper use of a table for information which should appear in the text.

CHANGES IN THE BLOOD observed in HAEMOLYTIC DISEASE OF THE NEWBORN

Estimation	Result
Haemoglobin	Lowered
Serum bilirubin	Raised
Reticulocytes	Increased
Direct antiglobulin	Positive

in any case should not duplicate the information. On the other hand a table is not the correct format in which to display data or information which can be given in a sentence or two of text (Table 1): that is just pretentiousness and most editors will disallow it.

When laying out a table prepare a number of drafts to get the most effective appearance. You should find the most economically spaced and well-balanced headings for the columns and make sure that columns of figures are properly aligned—whole figures aligned to the right and those with decimal places (including numbers below one, which should be prefixed 0.) having the decimal points in line. Try to avoid lengthy 'headings' for horizontal rows. When a whole column (or row) consists of the same figure repeated, omit that column and refer to the result in a footnote. Do be careful to distinguish between figures and symbols in the body of the table indicating a result of zero (0); no test performed (— or NT); no result obtained (ND = no data or not detectable).

Table 2 demonstrates many of the faults which people put into tables. The common factor (according to the legend) is a raised bilirubin level but the tests are not placed in any particular order to illustrate this; the 'Patient' numbers (actually sample numbers) should be aligned to the right, not the left; there is no need for a zero before a whole number (as with 09876 and 04987); the figures in the second column should be aligned on the decimal points and whole numbers should be followed by .0 (to show that they are whole numbers); the figures in the third column should be aligned to the right, not the left; the fourth column is unnecessary—if there are no results for any sample say so in a footnote; in the fifth column a dash is not the correct way to indicate a negative result and the strength of any quantifiable reaction should be indicated (as in the two preceding columns); the units of measurement should be indicated in the headings to columns two and three; the heading to column two is unnecessarily lengthy; the heading to column five is an undefined abbreviation which is not internationally recognized; and finally, the whole table is incorrectly boxed-in with vertical lines. The same data are presented in the correct form in Table 3.

Finally, double-check both the accuracy of all the data in each table and the necessity for each table in the context of your paper.

Table 2. Blood test results on 10 infants with hyperbilirubinaemia (NT = no test performed). (Example of how not to lay-out a table).

PATIENT	HAEMOGLOBIN LEVEL	SERUM BILIRUBIN	RETICULOCYTES	DCT
125879/B	12.5	75	NT	—
146320	13	83	NT	—
09876	9.2	308	NT	Pos
125880	11.7	71	NT	—
169417	12.5	76	NT	—
23807	12	65	NT	—
153216	10.9	68	NT	—
04987	11.4	70	NT	—
142310/A	13.1	84	NT	Pos
153215	10.9	187	NT	Pos

Table 3. Investigation of 10 infants with hyperbilirubinaemia (reticulocyte counts not performed). (Example of correct way to lay-out a table).

Sample no	Hb g/dl	Bilirubin μmol/l	Direct antiglobulin
23807	12.0	65	Neg
153216	10.9	68	Neg
4987	11.4	70	Neg
125880	11.7	71	Neg
125879/B	12.5	75	Neg
169417	12.5	76	Neg
146320	13.0	83	Neg
142310/A	13.1	84	+
153215	10.9	187	+++
9876	9.2	308	+++

Identify your illustrations

All your illustrative material will need to be identified in two ways; by numbering and by a title (or legend). All half-tone and line illustrations should be numbered in a single sequence as they are first referred to in the text. Use Arabic numerals, not Roman, and mark each item on the back very lightly in pencil or by a self-adhesive label. You should then indicate the approximate position of the figure in the text by writing in the margin 'Fig. 1', etc. with a horizontal arrow to show where you feel it should go. There are two things to note about this. The first is that each item in this class of material is known as a figure. Do not refer to photographs as 'Plates': this is a term used to describe separate pages of photographs, usually in a book, which are printed on a different (glossy) paper and generally 'tipped-in' to the binding separately with adhesive. The second point is that you cannot specify precisely where a Figure should appear due to the possible intervention of the end of a printed page and the variations in size of printed items from the originals. The positions which you indicate can only be approximate. Tables are dealt with in the same way as Figures only using a separate sequence of numbers (again starting with 'Table 1').

You will also need for each figure and each table a legend—that is, a short title which describes the contents of the item without including anything not to be found in the main text, but which can be understood by a reader without reference to that text. Legends should normally be typed on separate sheets (although each sheet can carry more than one legend) and these should be attached to the back of the finished typescript.

Copyright

Although copyright law is extremely complicated you cannot afford to ignore it if you intend either to quote anything written by anyone else or wish to reproduce someone else's illustrations. Copyright is the sole legal right to use of works created by an individual: it is not a right in the novelty of ideas. For all practical purposes science writers will encounter copyright only in respect of written works—books, articles in journals, etc.—drawings and photographs. If you wish to reproduce any copyright material the only safe procedure is to obtain written permission from the copyright holder—which means that you must appreciate at least the outlines of copyright law, which in Britain is subject to the *Copyright Act 1956* and the *Copyright (Amendment) Act 1971*.

In general terms copyright exists for a period of 50 years after the death of the author for literary works and for drawings and for 50 years from the end of the calendar year in which a photograph is first published. For all copyright material there are restrictions on anyone but the copyright holder reproducing the work in any material form, publishing it or (in the case of written works) making an adaptation of it. The general exceptions to the rule are described as 'fair dealing' for purposes of research or private study, or for criticism or review provided there is a 'sufficient acknowledgement' of the source. Specific exceptions can only be granted by the copyright holder who is the author of the work, unless it was made in the course of employment, when the employer will own the copyright, or in the case of a photograph, when it belongs to the owner (at the time) of the film upon which the photograph was taken.

A general exception to the author's ownership of copyright is becoming more common. Authors of journal articles are often asked to assign copyright to the publishers. This is to help with difficulties experienced in America with widespread abuse of photocopying facilities, but the effect is that it is the journal publisher who then holds the copyright and not the author. The fact is easily checked by looking for the printed symbol © which is followed by the name of the copyright holder and the year of publication.

Copyright laws in countries other than the UK are different—and that in the USA is especially complex—but there are international conventions and agreements which effectively ensure an interchange of copyright arrangements between countries.

Don't be tempted to follow the first Duke of Wellington's advice to 'publish and be damned'. Copyright can be complex—but if you are quoting, make sure you acknowledge your source if it is brief: otherwise get written consent. That way you will avoid trouble.

Chapter 5
Some other kinds of writing

> Ye see how large a letter I have written unto you with mine own hand.
>
> Epistle of Paul to the Galatians, ii, 9

> This shows how much easier it is to be critical than to be correct.
>
> Benjamin Disraeli
> Speech, House of Commons
> 24 January 1860

In addition to original articles there are a number of other kinds of written work with which you may become involved with a view to publication, and this seems a convenient place to consider them.

Short communications

Under a number of headings most science journals publish what might be called 'mini-papers' on topics such as modifications to apparatus or techniques and preliminary observations of phenomena which may be the subjects of subsequent full-scale research programmes and—in the fullness of time—original articles reporting them. Whatever the name given to this type of article you will do well to study both your target journal's instructions to authors, and also some recent copies of the journal, in order to gain an idea of the sort of thing which it includes under the heading. As a general rule, if a journal distinguishes between original articles and what its editor may call Short Communications, Short Technical Notes, Brief Reports, Preliminary Reports or what have you, then the distinction is often the basic one between research on the one hand and development on the other. Before you start, be sure which category your own work is likely to fit into.

When you come to write your 'Short Communication' you will have to see what style is prescribed by your target journal. It is likely that you will be expected to produce a short piece (maybe 800–1000 words) which is written in continuous prose without the IMRAD sub-headings of an original article. Nevertheless you will be wise to follow the principles of that system, covering the various elements of what you are attempting to convey in the logical order of why? how? with what results? and what does it mean (or matter)? The need to provide accurate references applies as much here as anywhere, of course.

Short papers are not necessarily—or even ever—easier to write than full articles. In the seventeenth century Blaise Pascal once apologized for having written a long letter by explaining that he didn't have time to make it shorter. It is a fact that the length of a written work is in inverse proportion to the ease of writing it. It is much simpler to be prodigal with words than to economize with them. You will find that if you are to remain within the stated length of a 'short' paper you will almost certainly have to do some drastic surgery on your first draft. Don't despair—it is good practice in communicating succintly.

Letters to the Editor

Most science journals carry a 'Letters to the Editor' section and this is often either an under-used or a mis-used section of the publication. Editors are generally well aware of this and so are correspondingly pleased to receive well-prepared and 'meaty' contributions. The functions of a 'Letters' section can be many but generally it provides a place for workers to comment upon, criticize, correct or supplement the contents of papers previously published in the journal. It may also serve as an alternative to a 'Short Communications' section for the rapid publication of preliminary observations which will be followed up by a full paper later. When a letter is received criticizing a paper already published in the journal it is usual for the editor to show a copy of it to the author whose work is under discussion and for him to invite a reply to be published simultaneously with the criticism.

The style of the letter is fairly open and will depend upon what it is you have to say. If you are referring to someone else's paper then give a full and accurate reference to it and do not waste words in verbal sparring. A letter beginning 'it was interesting to read Dr

John Bull's letter last month about his recent experiments on the incidence of overfeeding of domestic gerbals. This subject interests me very much. I have been working on the same subject for some time and I think that my figures are more reliable than his' is pure waffle. It would be better phrased more formally and briefly thus: 'Bull[1] has reported that x% of domestic gerbals are overfed. My own results do not agree with this.'

One mistake you should never make is to attack another worker personally in writing. This is a very regrettable practice in the arts and historians (in particular) can wax very vituperatively about their professional colleagues and competitors. Indeed, it is not unknown for libel actions to be pursued at law over alleged character defamation in published criticisms of one another's work. In the sciences we have fortunately been free of such unpleasantness and editors intend to keep it that way. If you must criticize, do so objectively and with good manners.

The form of a 'Letter to the Editor' will vary from journal to journal but you would do well to study the layout in a recent issue and copy it. For the most part editors find the use of headed paper from the writer's institution a nuisance as they either have to have the letter re-typed for the printer or send him something covered in a mass of red pen marks indicating what information is to go where or is to be deleted. Write—and have typed—your letter in the correct layout on plain A4 sized paper with double spacing and wide margins, with a proper heading above the commencement and your name and address in the appropriate places. Do not forget to sign it, however, (as proof of its origin) with the signatures of all the subscribers if there are more than one. Most editors will also thank you for the courtesy of a second copy of your letter so that one can remain in their files while one can be marked-up for the printer.

Finally, don't overdo your contribution to 'the literature' in this form. Some scientists seem to wish to make their names known by appearing frequently in print as authors of 'Letters' to the editors of several journals. If you really have something to say then say it: if not, don't risk acquiring the reputation of a sniper who is forever criticizing the work of others while producing nothing new yourself.

Book reviews

'One of the diseases of this age is the multiplicity of books; they doth so overcharge the world that it is not able to digest the abundance of idle matter that is every day hatched and brought forth into the world'.[1] These words, written over three and a half centuries ago, may be thought equally valid today. The editor of the world's first scientific journal, *Le Journal des sçavans*, recognized the problem in his preface to the first issue in January 1665. He stated that the journal would comprise 'Firstly, an accurate catalogue of the main books published in Europe': and he would not be content merely to list the titles but would describe their contents and assess their usefulness. These aims were also broadly followed in the first English language scientific journal, the *Philosophical Transactions of the Royal Society*, when it appeared four months later—although the reviews here were more descriptive than critical—and they have formed part of the editorial policy of most scientific journals since.

Although book reviews can be merely descriptive the value of such a review is mainly in making the book's existence known to a selected audience, and indicating its scope. The most valuable reviews are evaluative, the reviewer giving a personal assessment as to whether he considers the book worth reading or buying by the audience for whom he is writing. In essence this is an extension of the system of refereeing applied to papers submitted for publication and known in academic jargon as 'peer review'. The reviewer (like the referee) may be right or wrong in his opinions but he is giving the honest viewpoint of an independent, informed, and (it is to be hoped) coherent reader—which may be balanced against the publisher's sales 'blurb', and maybe the author's reputation. However, because a book review is the opinion of one man or woman (whereas several referees' opinions on a submitted paper can be assessed by an editor before making a decision) the selection of reviewers by the editor is critical—and good reviewers are hard to find.

Because scientific books are now so numerous (and sometimes so expensive) the critical review is often a vital factor in determining sales of new titles, as well as new editions of established titles. It is for this reason that unsolicited reviews (which may be influenced by hidden financial considerations) and unsigned reviews are not popular with editors. It is true that reviewers who remain anonymous can freely write critically about their friends, enemies or colleagues, but a signed review leaves no doubt that it is independent of publisher and author, who may each have an axe to grind under the cloak of anonymity.

So if you are invited to write a book review, how should you set about it? Surprisingly the first thing you do is not necessarily start to read the book from beginning to end. Begin by studying closely the title, the jacket 'blurb', the list of contents, any Foreword or Preface and any sales literature describing the book and its prospective audience. When you feel you have a general view of what the book is about, what level it aims at and for whom it is intended, it is time to browse through the book to gain an overall impression of the quantity and quality of the illustrations, the standard of printing and binding and the style of writing. Try dipping into the book at odd places to see what it has to say about topics in which you are particularly interested or knowledgeable: from this you will begin to build up an impression of the quality of the book. Finally, settle down to read right through the book if it is a reasonably-sized one, or in the case of very large volumes, at least a number of substantial sections of it.

When you come to write the review remember that you are in the business of objective evaluation, not subjective nit-picking. Begin by writing out the full bibliographical details. These are usually given in a particular order and while this may vary from journal to journal the following is that conventionally used in most publications:

1 Full title of the book in lower case letters with capital initial letters (except for definite and indefinite articles and conjunctions), all underlined.

2 Followed by:
the edition (if other than the first)
the authors' (or editors') names, preceded by
their initials. In the case of editors follow the names by '(Editor)'
the place of publication, followed by a colon
the name of the publisher
the year of publication

the ISBN number (usually found near the © notice)
the number of pages
the price

3 Followed, starting on a fresh line, by the review which should normally not exceed 300 words unless this has been requested or the book is of exceptional importance.

The review itself might well start with a (very) short description of the contents of the book but should be mainly a personal assessment of its worth in the context of the intended audience, the price range and other books available on the same subject. Try to be relaxed in your writing style and don't fall into either of the two common traps for the inexperienced reviewer: don't fill the review with examples of specific errors—be they the author's or the printer's—and don't lay stress on what you may feel to be a high price, unless you can demonstrate that it is quite exceptional by comparison with that of similar books on the same subject and of similar size. If both hardback and paperback editions of the book have been published draw attention to the fact and quote both prices. Finally, as with all other forms of writing, put your review in a drawer for a week or two before finalizing it and sending it off.

In most reputable journals book reviews are signed by the reviewer who thus takes personal responsibility for what is said—be it praise or condemnation. It should scarcely be necessary to mention the law of libel—the editor will certainly jump hard on anything you write which might be defamatory—but neither should you go to the other extreme of sycophancy in the hope of currying favour with an eminent colleague. What is called for is honesty, a balanced professional judgement and total objectivity. Writing a good readable book review is neither easy nor a task to be undertaken in odd moments. It is a lengthy and demanding job which calls for practised skill in writing in a telling manner but in very few words. If you are asked to review a book feel honoured, but be prepared for some hard work.

Case reports

Although they are usually considered essentially clinical in nature, case reports and nursing care studies of patients whose disease process is unusual, or for some reason of particular interest, often contain a substantial amount of scientific data and a scientist may be a co-author of such a report. Indeed, clinical medicine is itself sometimes regarded as a science in its own right.

In writing about people there are certain points which must be taken into account whatever the style of the report itself. The most important of these is that the identity of the patient is never revealed. Coming a close second is the corollary that one does not de-personalize any individual by referring to them as 'the case' or 'the person *that* ...'. These two requirements can lead to some clumsy circumlocutions in referring to the individual and a number of ploys are available to resolve the problem. In reports in medical journals it is usual either to refer to the patient by his or her initials (e.g. 'SR, a woman of 25 years 3 months, was admitted to hospital ...'), or by a general description (e.g. 'A man aged 32 years was treated ...'). In nursing case reports it is more common to use the patient's first name only (e.g. 'Catherine was married, aged 28 and pregnant for the first time ...'). Any of these methods is acceptable and you should follow the convention of your target journal.

Related to the problem of patient identity in the text is that of identification in photographs. In any photograph of the patient the face must be masked to hide the identity of the individual unless there is written informed consent by the patient or his/her legal guardian for its publication untouched. Unless facial appearance is crucial to the report there should normally be no need to take this latter course.

The format and writing style of case reports varies quite markedly between medical and nursing reports. In medical case reports it is usual to write in a strictly factual style, listing the results of physiological and laboratory investigations and drawing conclusions from these. The reports are usually presented in three sections: an Introduction, the Case Report, and a Comment or Discussion, followed by any references.

In nursing reports a less formal style is often adopted which makes easier reading but which also gives a distinctly un-scientific tone to the report. Members of a 'caring' profession will justify this but it is easy for the informal style to become unacceptable literature. Statements like 'When Mary's observations were taken they were written down' are neither grammatically correct nor do they aid clarity. Similarly, servile references (without the definite article) to (e.g.) 'Doctor was called' are representative of an outmoded style of address and of prose. Probably the most common fault with writing in this type of report is failure to exercise self-criticism and to edit out the 'waffle' which creeps unwanted into all of our writing if we let it.

In writing case reports for any audience you should aim at a relaxed style of writing combined with an economy of words and precision of phrasing. It sounds complicated and it really isn't as easy as it seems. The sort of style one would use for a 'Short Communication' (or even a 'Letter to the Editor') in a science journal is probably better than that which one often sees in printed reports. Most people don't take this type of writing seriously enough; and all too often it shows.

A word about writing for foreign journals

It may be that you will wish to try your luck with a journal published in another country. If you are writing in English this will mean considering differences in vocabulary and spelling between (British) English and 'American' (English). As a general rule most Commonwealth and European journals published in English follow British usage, except for Canadian journals which tend to use American vocabulary (e.g. *elevator* rather than *lift*) but English spelling (e.g. *technique* rather than *technic*). Faced with this knowledge you should always aim to use the vocabulary and spelling which are conventional in the country where your target journal is published.

Differences in vocabulary between English and 'American' are actually minimal as far as science writing is concerned, and even in everyday speech there are not a large number of differences. Apart from such well-known examples as American 'side-walk' and 'gotten' for English 'pavement' and 'got' most of the more obvious examples seem to be connected with motoring—for example the English 'bonnet', 'boot', 'silencer' and 'accelerator' become 'hood', 'trunk', 'muffler' and 'gas pedal' in America. In medicine and science the only likely source of difficulty is that certain words which in Britain are generic (e.g. 'adrenaline') are registered trade marks in America for manufacturers' specific products (e.g. 'Adrenalin', for which the American generic name is 'epinephrine').

Spelling is a more obvious point of difference in scientific language on the two sides of the Atlantic. Probably the most common—and from the English point of view most readily recognized—difference is in words which in English possess a diphthong. American usage does not recognize this compound vowel form, resulting in spelling differences such as:

English	American
diarrhoea	diarrhea
foetus	fetus
haematoma	hematoma
haemoglobin	hemoglobin
haemorrhoids	hemorrhoids
oedema	edema
oesophagus	esophagus
oestrogen	estrogen

Other than this category of spelling differences there are some general rules which can be followed. Shown in Table 4 are some examples.

Table 4. Some examples of spelling differences between English and 'American'.

In English	e.g.	In American	e.g.
Nouns ending in -our	*Colour*	end in -or	*Color*
Nouns ending in -re	*Centre*	end in -er	*Center*
Words with double consonants	*Programme*	have single consonants	*Program*
Verbs ending in -ise	*Liquidise*	end in -ize	*Liquidize*
Words ending in -que	*Technique*	end in -c	*Technic*

Do note, however, that when you are quoting, or citing references to, published work you should always use the spelling of the original and not seek to alter it to that of the rest of your paper.

Few scientists whose native language is English or 'American' will wish to prepare papers in any other language, and this is certainly not a recommended ploy for beginners in science writing.

Suffice it to say that if you are fluent in scientific French, German, Japanese or Russian and do intend to tackle a paper in one of these or a similar language you should aim to write it in English first and then prepare a translation. My own advice, however, would be for you to gain experience in writing in English for some time before thinking of branching out in any other tongue.

And what next?

So now you have finished writing your article, what do you have to do to see it in print? That is what we shall consider in the next chapter.

Chapter 6
How to prepare and submit your paper, and deal with editors and printers

What I have written I have written.
St John's Gospel, 19, 22

Now Barabbas was a publisher.
Attributed to
Thomas Campbell (1777–1844)

It may seem unnecessary to say it but having written your paper you will need to have it typed. As an editor I have received submissions in all forms, from handwritten scripts on lined paper to printed versions of the article which had already appeared elsewhere. Be advised: anything other than a properly prepared typescript of something which is original and unpublished will almost certainly be sent straight back to you, unread.

Most journals require you to submit two or three copies of your paper, one of which must be the original typescript. While carbon copies can be sent they tend to be pale, indistinct and inferior to the top copy. It is better to use xerographic copies (which are not strictly the same as photocopies, which have a 'wax' surface and are little better than carbons). The reasons for submitting multiple copies are to expedite the handling of your paper and to protect the valuable original.

When handing your paper to a typist do make sure it is accompanied by a copy of the 'Instructions to Authors' of your target journal and a request that these be followed closely. No two journals adopt quite the same style and layout for their papers and this can cause much grief for all concerned. Authors whose papers are rejected by one journal have to have their papers re-typed for submission to other journals, while editors are faced with a lot of tedious (and unnecessary—not to mention expensive) work thrashing a paper into shape if it is not in the correct style. In an attempt to rationalize this chaos (at least partly) the editors of over 150 journals in the biomedical sciences have agreed to receive manuscripts prepared in accordance with a set of uniform requirements prepared first in 1978 in Vancouver by an International Committee of Medical Journal Editors and known informally as 'the Vancouver style'. This does not mean that all these journals will henceforth publish everything in the same style but that if a typescript is prepared in the specified style the editor, rather than the author, will make any necessary alterations to make it conform to the particular journal's house style. If you are intending to submit your paper to one of the journals whose 'Instructions to Authors' refer to the 'Uniform requirements for manuscripts submitted to biomedical journals' then get a copy of these requirements[1] and hand it to your typist.

Because the uniform requirements are general to a number of journals there will be some additional requirements peculiar to each journal—the number of copies to be submitted, which languages are acceptable, types of paper which will be considered, etc. Do take these into consideration also. In particular note that the uniform requirements refer to the system of citing references used by *Index Medicus*. This style does not use capital letters for the spelling of authors' names (as is done in the British Standard 1629 :1950) so don't you do it either. Editors would far rather—and can with far less fuss—alter lower case type to capitals for the printer than *vice-versa*.

Other points—with which an experienced typist should be familiar but which you may have to explain to a less experienced one—are as follows:

1 The paper must be typed in double spacing with wide margins on one side only of plain A4 size paper and every page must be numbered.

2 Words are only underlined if they are to be printed in italics (as with the names of bacteria, the names of journals and the titles of books) and underlined with a wavy line if they are to be printed in bold type.

3 Words should not be broken at the end of a line.

4 Numbers for reference citations in the text should be typed as superscripts (i.e. raised by half a space) and not placed within brackets.

5 Tables should not have vertical lines drawn in them.

6 Numbers of less than ten are spelled out in full and numerals only used for larger numbers when specifying quantity.
7 Numbers with five or more digits a space, not a comma, is used between each set of three figures. (This is because in continental usage a comma is used to represent a decimal point, and this provides too much scope for confusion. The figure '3,000' means 'three thousand' in English but 3.000 to a European.)

When the typescript is complete the first thing to do is check it over for any typing errors (not every typist is clairvoyant and not every scientist's handwriting is copper-plate) and then for consistency in such matters as units of measurement (I recently received a paper which referred indiscriminately to solutions as g/l, %w/v and containing '*x* moles': stick to one type of measurement throughout), definitions of abbreviations, correct use of nomenclature for such technicalities as bacterial names, blood group symbols, chemical and biochemical compounds, radioisotopes, etc.

Submitting your paper

By this stage you should have a good idea at which journal you are aiming—your 'target journal'. If you have not, then give the matter some serious thought. Presumably you are writing in order to spread the word of your discoveries. You will not do this if you do not take account of the size and nature of the readership of your target journal. A paper on a clinical topic will be wasted on the readers of a laboratory-orientated journal just as a paper on pure chemistry would be wasted on the readers of a journal devoted to orthopaedic surgery. Similarly, if your paper is on a topic which has multi- or inter-disciplinary appeal a multi-disciplinary journal will be better than one which specializes in a rather narrow field. Most scientists know the literature of their own subject interests, however, and should recognize almost instinctively where their paper will make the most impact.

Having selected your journal, and presuming that you have had no drastic second thoughts after the typescript has rested in a drawer for a couple of weeks (see Chapter 2), then you can make it ready for sending off. Be sure that you have the number of copies specified in the 'Instructions to Authors' (including a set of illustrations with each copy, held together in a separate envelope and not with staples or paper clips which can irretrievably ruin photographs or drawings). Whatever you do, keep a further copy of the paper (complete with illustrations) yourself: postal services are not always as reliable as they once were.

You will need to include in the package a covering letter. Again, it may seem obvious but do remember to say in this letter that you are submitting the paper for possible publication. I have too often received envelopes containing a letter saying merely 'Please find enclosed three copies of my paper. Yours sincerely'. The article—and its copyright—are your property and no editor likes to be 'piggy-in-the-middle', holding a typescript with no authority to do anything with it.

Most journals will require a statement, to be signed by all the authors of a paper, to the effect that the work is original and that it has not been accepted for publication nor is concurrently under consideration elsewhere without permission of the journal's editor and/or publisher. All of this is mainly to avoid wasteful (and academically unacceptable) dual publication, but it also ensures that no-one has their name put to something they don't know about. Increasingly, journals are also requiring signed statements from all named authors of multi-author papers that each of them took part in the planning, execution or analysis of the work reported and/or the actual writing of the paper (see Chapter 3). You must also include signed letters of permission from the performers of any hitherto unpublished work and the owners of any copyright material you have included in your paper (such as previously published diagrams, etc.). If your work included the use of people or live animals you will also need a statement from the appropriate ethical committee or holder of the relevant animal licence, that the work was carried out according to the normally accepted rules.

Finally, if there is more than one author do not forget to say to whom the subsequent correspondence should be addressed. You may then pack up the scripts, illustrations and covering letter in a stout envelope, with card stiffening if there are illustrations enclosed, and send it off to the address given in the 'Instructions to Authors'.

What happens next?

Having posted your *magnum opus* do not expect too much to happen too soon. Within a week to ten days you should receive an acknowledgement of receipt of your paper and then you can expect a wait of several weeks before you hear any more. All reputable academic periodicals—including science journals—operate a system of peer-review by which every paper submitted is seen and criticized by experts in that particular topic. Usually two, and sometimes three, referees see each paper; and although the editor's decision on each paper is entirely his own—and usually final—the opinions of referees carry great weight.

It is usually at this point that some of the inevitable delay in handling papers occurs. If two referees are in fundamental disagreement over a paper the editor will usually seek a further opinion. Many editors, and virtually all referees, fulfill these functions on an honorary part-time basis and in their full-time posts as scientists are very busy people. Not for nothing is it said that if you want a difficult job done you give it to a busy man. Notwithstanding their other professional commitments, family lives and even holidays, most referees still manage to produce perceptive and constructive reviews of the papers sent to them within the time limits set by editors.

It is worth pointing out that referees work under strict constraints of confidentiality. Research work, and reports of it which are submitted for publication, are the property of the author(s) and it is understood that editors and referees are in a privileged position in having access to this unpublished information. They may not make use themselves of anything which they learn from this situation, prior to actual publication of the paper.

Once the editor has received the reports of his referees he assesses the paper in the light of them, makes his decision and writes to the author—who may expect one of four responses.

(i) If you are very lucky indeed you may be told that your paper has been accepted. However, very few papers are accepted without at least some modification. If you have struck lucky you can congratulate yourself: if not then you have the consolation of being one of the majority (98–99%) of science authors in this situation.

(ii) If your paper is a good one, and the work which it reports is sound, it may be accepted subject to minor changes. These may be, for example, a request for further details about some point, clarification of an ambiguous or obscure passage, or suggestions for additional tables or figures. You will rarely be asked to improve small points of grammar or style: this will normally be done by the editor (usually without further reference to you) without the sense of the text being changed. Your response should be to comply with the editor's requests as quickly as possible and again congratulate yourself on having got off lightly.

(iii) Most papers which are not simply rejected will be accepted subject to (more or less) major changes. If your paper falls into this category you will either be given copies of the referees' reports on it or these will be quoted in a (lengthy) letter from the editor. This is the occasion when your Achilles' heel is uncovered and any weakness in planning, execution or analysis of your work is exposed. Referees, like examiners, are basically nice men and women who are interested only in maintaining high academic standards. If they see a weakness or a fallacy they will point it out because that is why they are there; not because they dislike the sound of you or your laboratory. The important point is that you swallow your pride and consider objectively whether you do or do not agree. For the most part you will be obliged—if you are honest—to admit some substance to the criticisms. Again, be thankful. If you did not have a chance to meet—and counter—these criticisms privately you would certainly have to meet them (probably in public) from many more equally perceptive critics if your paper were published unaltered.

When considering referee's criticisms you are not necessarily expected to swallow hard and take everything on board without argument. If you believe a referee is wrong or has misunderstood what you were trying to say, then write back to the editor to say so and to explain why. First, however, consider whether the fact that an experienced referee has misunderstood you is not an indication that you have not expressed yourself clearly, and that most of the 17 000 readers of the journal may misunderstand you similarly. Sometimes all that is needed is a radical re-writing of a section.

When you believe you have met—or defended—all the criticisms in the referees' reports re-submit your revised paper. At this stage do not fail to quote any previous reference number which the editor may have allotted to your paper, and enclose (on a separate sheet of paper) a list of all the places in the paper where you have made changes and a note of the nature of these changes. This will enable the editor and referees to see whether or not you have satisfactorily met the points which they raised.

As part of this exercise you may also be requested to change the title of your paper or to have parts of it (especially your list of references) re-typed in conformity with the journal's requirements. The lesson here is to get it all right in the first place (see Chapters 3 and 4).

Above all, please do not start an acrimonious correspondence with the editor. He has nothing personal against you and even if you feel the righteous indignation of a lioness defending her cubs when you see your paper being criticized, try to realize that you and the editor both have the same objective—a high standard of academically respectable publication which will bring credit upon both author and journal. If you disagree with the referees either justify your disagreement objectively or withdraw your paper and try it on another editor. If you can meet the points made you will be doing yourself a favour and probably learning something further in the process: and that can't be bad.

(iv) More than 50% of papers submitted to reputable science journals are rejected. This is not a target figure which editors aim at but represents the result of a careful application of quality assurance. When you receive a rejection note it may or may not be accompanied by reasons or by the referees' reports. Do read the wording of the letter carefully, however. Few of us can rise to the sublime heights of the (possibly apocryphal) rejection letter said to have been sent from an oriental journal:

> 'Sir. We have read your manuscript with boundless delight. If we were to publish your paper it would be impossible for us to publish any work of a lower standard. As it is unthinkable that, in the next thousand years, we shall see its equal, we are, to our regret, compelled to return your divine composition. We beg you a thousand times to overlook our short sight and timidity'.

Surely the epitome of the *double entendre*.

In one way a normal rejection letter may be indicative, however. If the editor says that your paper 'is not suitable for' or 'would not fit happily into' the journal do not waste time seeking permission to re-submit your paper. Either you really have picked the wrong journal or the editor may be using the phrase as a kindly euphemism for saying that either the work or the paper (or both) are simply not up to his standards.

If you have not been sent copies of the referees' reports upon which his decision was based it is probably because the editor—from bitter experience—knows that few authors like criticism of their papers any more than most car owners like criticism of their driving, and many use it as an excuse for a heated (and pointless) correspondence which serves no purpose and only profits the postal service. If you really want to see the criticisms, and are prepared to swallow your pride and accept them, most editors will supply them on request. Having read the reports, do not modify your paper and re-submit it without seeking permission of the editor first. If his 'In' tray is particularly full he may be unable to tackle a marginal submission and you would be better advised to submit it to another journal: but never submit any paper if it is still under consideration elsewhere.

Checking proofs—and afterwards

If your paper has been accepted there will be another delay of several weeks (although this time-scale may vary with the frequency of publication of the journal) until one day you find an envelope in your mail containing proofs. Proofs are the printer's first setting of your paper and the nature of them will vary. Traditionally printers used to provide long strips of paper the width of a column of print in the journal and known as 'galley proofs'. Some journals still provide these but most printers now use a different style of typesetting and go directly to 'page proofs' which, as the name implies, are layouts in the style of the printed pages of the journal. These are generally set on a special electronic typewriter and reproduced photographically to make plates for the printing machine. In either case you will be expected to correct the proofs.

It is important that you realize that at this stage correction is all that you do. It is too late now to have second thoughts about the content of your paper. It is errors in typesetting for which you are looking, not comments which you have thought up since the paper was accepted. Indeed, most publishers are prepared to charge authors for the cost of re-setting alterations in substance, or additions, made at the proof stage.

Once you have got over the glow of pride at seeing your work—and your name—in print you must get down to some careful work. Generally you will be supplied with either the original or a xerographic copy of the marked-up typescript which was sent to the printer. With both this and the proof in front of you go through the paper word by word looking especially for mis-spellings, transposed letters, missing words, numerical errors in tables and calculations and incorrect initials with authors' names in the references.

For every error which you find you will need to make two marks—one on the actual error and a corresponding mark in the margin beside it—using a colour of pen different from any which may have been used already by the printer or the editor. Some of the more common British marks are shown in the Appendix but do not worry yourself sick about these. Proof correction marks differ between British and American printers, while European printers do not exactly agree with either system. If you do not know the correct symbols and the printer has not supplied you with a list of them just ring around the error in the text and clearly write the correction in the margin beside it.

Failure to correct every trivial error can lead to comical results, as in the new edition of a text-book which I saw recently in which the 'Acknowledgements' noted that 'The authors would like to express their gratitude to colleagues who have *impaired* enlightenment and assistance'. I hope that the original typescript said 'imparted'. More seriously, uncorrected typographical errors can lead to tragedy. A colleague who was reviewing a book for me a couple of years ago discovered a displaced decimal point in a figure giving the dosage of a drug. If the printed figure had been followed the results would have proved fatal to the recipients and prompt action caused the book to be withdrawn and copies recalled for correction of the error—which should have been detected at proof-reading.

When you think you have corrected all the errors of typesetting read right through the paper and see how it comes across to you: you may even notice further errors at this stage. Do not worry about the quality of half-tone illustrations as proof copies are never to the same standard as the final version; neither should you take any notice of strange sequences of letters and/or numbers at the top or bottom of any sheet—they are solely for the printer's benefit and will be removed before final publication.

When your corrections have been completed you should return the proofs—and the typescript if it was sent to you—to the address indicated. You may also be asked to complete a form if you wish to receive offprints of your paper.

Do not be disturbed at this stage at what will probably seem to you to be an inordinate delay in the appearance of the paper. Authors are often distressed at the length of time which usually elapses between sending a typescript to a journal and eventual publication. No editor likes undue delays—and some weekly or monthly publications (e.g. the *British Medical Journal*) make provision for very rapid publication of unrefereed short items of 100–150 words—but generally delay is inevitable. Many journals appear quarterly and nearly all have a printing schedule as long as the interval between issues. This means that a paper accepted in mid-January may not appear before the July issue. Before acceptance, however, the refereeing process normally takes up to six weeks and if a paper has to be returned to the author for alteration (as is the usual case) a further delay is introduced, probably compounded by the subsequent need to obtain further comments from the referees. On top of this, many journals have more material for each issue than they can print so papers must go into a queue for publication. On balance you should expect at least seven to ten months from submission to publication. Up to two years is not unknown.

Offprints and reprints

'Offprints' and 'reprints' are not the same thing, although few people seem to appreciate the difference and they fulfill the same function. Essentially these are printed copies of your paper. When the great day eventually comes that a copy of the journal lands on your desk and you see your article in its final form, set amidst other work in your subject area, you will doubtless wish all your friends and (you hope, envious) colleagues to be aware of it and with this in mind copies of the paper are usually made available by the publisher. Offprints are further copies printed in the same run as the journal: reprints are made subsequently from the film or printing plate, if they have been retained, or a film is made using the printed journal. You will appreciate that offprints are much cheaper than reprints but this can only benefit you if you can predict your needs and order them in advance. Many journals will give you a small number free of charge in lieu of payment for your article.

The real demand for offprints will probably be clear to you only after your first paper has been published. Starting a month or so later, and probably lasting over six to twelve months you will receive a steady flow of requests for 'reprints' of your paper from all over the world. Most requests will come from North America and Western Europe but there will be many from developing countries and from behind the iron curtain. The reason for this is that although the journal in which your paper appeared probably has a circulation of only a few thousand copies, if it is an established and reputable journal it will be covered by abstracting and indexing services which are widely subscribed to by libraries and institutions which cannot afford to take all the individual journals they would like. Thus many of those requesting your paper will not have read more than its title: hence the importance of choosing that title very carefully indeed.

Whether you will wish to supply maybe a couple of hundred enquirers with 'free' literature at your own expense (and don't forget the postage costs also) will be up to you—especially as many American scientists routinely have their secretaries send out preprinted requests for copies of everything published in their speciality for their 'reprint library'. (Incidentally, the curt ill manners of some of these printed cards will surprise you. It is not uncommon for a 'request' to ask for three copies of your paper 'and of anything else you have published on the same subject'.) Many British workers compromise by sending offprints only to enquirers from the less wealthy developing countries, as a service to scientists who lack the library and financial resources of a developed country. Whatever you decide, don't say you haven't been warned of the problem.

So your paper has been published and become known to your contemporaries. When the dust has settled you will probably wish to write another one. You will find that it gets less difficult with practice, and you will get more self-critical. Few authors can go back and read their early publications with equanimity. If this is true for you (as it certainly is for me) then rejoice, for you are beginning to learn.

Chapter 7
How to write project reports, dissertations and theses

> If a man will begin with certainties, he shall end in doubts, but if he will be content to begin with doubts, he shall end in certainties.
>
> Francis Bacon (1561–1626)
> *The Advancement of Learning*

As we looked at essay writing back in Chapter 2 you may wonder why we have only now got to the point of considering other written forms of examination. The answer is that each of the types of written work we shall consider in this chapter is based upon the general format or writing principles of a scientific paper—and if you know how to write a paper then you are well on the way to knowing how to write a project report, dissertation or thesis. As these are all variations on the same theme—examinable written reports, of increasing complexity and sophistication—let us consider them in the order in which, as a scientist, you are likely to encounter them during your career.

Project reports

As part of many taught courses in science, especially for undergraduates, the student is required to undertake a laboratory-based project—essentially an investigation into a defined area of a discipline—but not at the level of original research. The report of this investigation has to be written up in a satisfactory form and is marked as part of the assessment process. The object of the exercise is not only to measure the student's practical ability but also to introduce him to the academic discipline of writing a clear report in the accepted formal style.

The shape of the project report is very much that of a conventional science paper, with an Introduction and sections devoted to Materials and methods, Results, a Discussion and a supporting list of references. As this will probably be your first attempt at this type of written work do give yourself plenty of time for writing. You will be well advised to start on the Introduction as soon as you have done sufficient preliminary reading to have the nature of the problem you are studying, and a plan of campaign, clear in your mind. Indeed, as you will be undertaking a completely new form of exercise you can hardly begin writing too soon as you will probably have to make several attempts at the first section you write before you get the 'feel' of this type of writing and start to be satisfied with what you produce.

If you have read the earlier chapters of this book you should by now have a fair idea of how to proceed. Aim to write each section as your practical work goes ahead. Once you have established your working procedures draft out your 'Materials and methods' section and when you start collecting results start to put them into a suitable order for writing up that section as soon as they are all to hand. By the time the last result is entered in your laboratory notebook you should have little more than the 'Discussion' to write, tables and diagrams to prepare and some tidying and polishing up to do.

As far as references are concerned you should build up a card index of these, entering one reference to a card and writing out each reference in full as you look it up or read it. Using the sequential numbering system and writing the draft of your paper as you go should make the numbering of the cards and the insertion of the numbers in the text easy. An alternative approach is to enter the names of the authors into the draft of your text in the style of the 'Harvard' system (see Chapter 5) and file your card alphabetically under the first authors' names. When your draft report is otherwise complete and ready for typing (or writing out a fair copy, if that is allowed) you can number the citations as you come to them (from the beginning), delete the names in the text and add the numbers to the cards which can then be re-sorted into numerical order.

Don't overlook the regulations laid down for presentation of your project report. Either write out a fair copy in legible handwriting or have it typed in the same way as a paper for publication (see Chapter 6). Most colleges permit an inexpensive form of ring or slide binder but whatever the regulations say—follow them.

If your project report is well-regarded by your tutors you may wish to submit it to a journal for publication. If so, do not submit it without very careful revision. The standards for publication are usually higher than those set by universities and colleges for components in the assessment of undergraduates. In particular, even if your report is well written and presented, the level of originality in the investigation may not merit publication. Do seek advice from your tutors about this and if you decide to go ahead take care to comply with the 'Instructions to Authors' of your target journal. In any case, good luck with your first serious writing job.

Dissertations

There are various definitions of a dissertation but that which I like best is the one which used to be used in the 1950s by the (then) Institute of Medical Laboratory Technology. It described a dissertation as 'an ordered and critical exposition of existing knowledge of a subject'. While in strict terms a dissertation may be made orally or in writing it is only the written form which is now used in (some) examination systems. Some people look upon a dissertation simply as a literature review. It has much in common with review articles published in science journals and the writing of such an article is not significantly different from the writing of a dissertation—except possibly in respect of any constraints on length.

This is one form of written work which does not follow the basic IMRAD format (see Chapter 3), there being a fair degree of freedom for the author to choose a format best suited to the subject matter. The first step in planning, indeed, must be to sketch a skeleton outline of proposed subject headings. You will have to decide these for yourself but every dissertation (or review article) will need a beginning (an Introduction) and an end (a section devoted either to a summary, a discussion or some conclusions): you will also need a fairly extensive list of references. What comes in the middle will vary with the subject but generally you should treat your subject in a logical sequence with as many (or as few) main section headings as seem desirable.

As with any other form of science writing, but with even greater care, you will need to take careful note of the precise details of each paper that you cite as a reference. There will be a lot of them in this form of writing and it would be too easy to lose track or to introduce bibliographical errors.

The style of your writing will need to reflect the style of your study of the (always primary) literature sources. The essence of a dissertation is not just that it is a review, but that it is a critical review. It would be too easy—and too boring and too useless—merely to list what other people have done: 'In 1968 Jones did that while in 1969 Smith did the other. A year later Brown (1970) did something else'. You will get neither credit nor praise for such writing. The correct approach is to select only the literature which is significant and critically to review each of the papers you select, comparing different workers' results, theories and conclusions and assessing each in terms of the others. In such a study it may well be difficult to reach any firm final conclusion other than the need for continuation of research in the subject. Do make a firm final ending for your dissertation, however.

Theses

A thesis is essentially a report of an investigation performed by the candidate, embodying original research, representing a definite contribution to knowledge of the subject, and submitted for examination for the award of a higher degree or professional qualification. In addition to reporting work done and drawing conclusions from it the presentation of the thesis is also seen as evidence of the candidate's ability to apply a disciplined and systematic approach to scientific investigation and to convey his findings in a logical, coherent and literate manner in accordance with accepted academic practice.

Research may legitimately be conducted as a group or team project, but work to be considered for submission as a thesis must be primarily the work of the candidate. Originality in identifying suitable areas for investigation is important and you should seek to produce a title and synopsis which reflect your own approach to a problem, rather than merely adopting a line of research because it complements—or even parallels—other ongoing work in your department. When an examining body accepts a programme which forms part of a joint or group project you must indicate clearly the extent of your individual contribution and that carried out in collaboration with others.

The Shorter Oxford English Dictionary defines a thesis as 'a proposition laid down or stated, especially as a theme to be discussed and proved, or to be maintained against attack'. You should bear this in mind and work towards a proposition which you are able and prepared to defend as examiners will want to know if your proposition is tenable; and by subsequent oral examination establish both its originality and your ability to defend it. Although the disquisition upon the written thesis is no longer made in public, nor conducted in Latin (as was at one time the convention in European universities) it remains a vital part of the examination of the thesis in establishing that the work reported really is your own, and that criticisms which may be made of any aspect of it are capable of refutation.

Generally speaking a thesis should be written in the style of a science paper prepared for publication, following the IMRAD format (see Chapters 3 and 4), but with three main differences. (i) The thesis will be much longer than a science paper, for (ii) it must be exhaustive rather than representative in reporting and commenting upon your experimental work. Also (iii) it will probably contain a more extended historical literature survey in the Introduction. As with a project report you should start writing very early during your laboratory work, and you will require to produce an extensive and accurate list of references.

Once the thesis is written in draft form you must take careful account of the regulations for presentation which your examining body uses. Many authorities require that these shall conform to the British Standard 4821–1972, *Recommendations for the Presentation of Theses.* The presentation of his or her thesis is something which almost every candidate takes too lightly and you should provide yourself with a copy of BS 4821 : 1972 and ensure that your thesis conforms exactly to the relevant requirements. Points to which you should pay particular attention include information provided on the title page, the scope and position of the Summary, the exact style of the reference citation system to be used and the position and content of the lettering upon the binding.

Ability to understand and comply exactly with instructions is implicit in the discipline necessary to undertake and report upon original work and a candidate who does not take the trouble to present his thesis strictly in accordance with the regulations is exhibiting a careless and sloppy approach to his work which reflects badly upon his ability to undertake the discipline of independent research. You will be expected to take as much care in ensuring that the presentation is accurate as in ensuring the accuracy of the work reported.

Examiners will study your thesis with particular regard to clear and logical expression, good grammar, accurate spelling and presentation, and will not hesitate to return a thesis for correction of any deficiency in any of these areas. An otherwise acceptable bound thesis which does not contain (for example) a Summary, or a title page conforming to the published requirements is likely to be returned to the candidate for the binding to be opened and the deficiency made good before any recommendation for a pass result can be made.

A biochemist friend of mine once submitted a lengthy thesis to a Scottish university for the degree of Ph.D. Following his oral examination he told me that his examiners would not recommend him for a pass until he had made nine alterations to his typed and bound thesis. He then showed me the list of required alterations: every one was a minor correction of grammar or punctuation. In the business of writing theses for higher degrees only one standard of writing is acceptable—perfection.

Chapter 8
So you fancy writing a book?

> Books must follow sciences, and not sciences books
>
> Francis Bacon (1561–1626)
> *Proposition touching amendment of laws*

> Of making many books there is no end
>
> Ecclesiastes, xii, 12

Sooner or later in many a scientist's career there comes over him a feeling that he should write a book. Often the trigger is the realization that there does not seem to be a suitable text book for a class of students in a particular discipline at a particular level (which is how my own first book came to be written), or perhaps in a newly developing subject there is no single comprehensive book at all. Before you decide that you will make your fame and fortune by supplying the missing tome, however, you should consider three points very carefully indeed. The first is that writing a book which is to have any chance of success is very hard and time-consuming work; the second is that very few people indeed achieve fame from writing a book—and even less make significant additional pocket money, let alone a fortune; and the third is that unless you are proposing to take a huge gamble with your time and energies there is a strong chance that you will never see your brain-child on the shelves of any book shop. Indeed, too many people fail because they choose the wrong subject or tackle it in the wrong way, or simply lack the right motivation.

One successful publisher has summarized the difficulties very comprehensively[1] and I can do no better than quote his very pertinent comments:

> Many of us have at least one book in us, but anyone with a vague wish to write without any information to impart to any particular audience would be better employed in reading—and living—until a subject and an audience can be defined. So many who have the germ of a book in them fail to work out

who the readers are likely to be (or whether they exist) and to orientate the work to please and interest them. The commonest mistakes in approach might be summarised thus:

(1) Being interested in writing only about what is already known personally or can be discovered easily, and thus failing to produce a viable subject coverage.

(2) Though coming closer to producing a viable subject coverage, marring it by giving unbalanced vent to some theory, love or hatred.

(3) Being unable to select a good balance of material and, for instance, floundering when trying to put a local or limited subject in broader perspective, or giving too much minute detail and not enough general picture.

(4) Skimping the job, to speed publication and financial return, especially by failing to complete research on certain aspects.

To this I would add failure to take seriously the business of how to write clearly, logically and grammatically. The same publisher went on to point out that 'Very many more books are planned than written; many more are started than finished; many more are finished than published; and many more are published than really succeed in their aim'[2].

Now all of this can be very depressing but still people do write books, have them published and see them succeed, so if you are certain that you have a book in you, struggling to get out—and that you can meet all of the objections—then you will proceed despite the gloom. Really, the critical question to ask yourself is 'would I spend my own money on such a book if it were to be published tomorrow?' If the honest answer is yes then you may indeed have identified a vacant slot in the market.

How to set about it

As with all writing there are two essential preliminaries, although with a book they are perhaps even more essential than usual. The first is to identify your market. Determine for what audience you are writing. Are they undergraduate students, research workers, qualified scientists engaged in routine work, or the lay public? Your answer will determine the form, approach, breadth and depth of your coverage. The second point is to plan your approach, in detail.

The IMRAD formula will not apply and you will need to consider the most logical and coherent order in which to present the subject matter. Indeed, the first two things you should write—no matter how sketchily or on what scrap of paper—are your title and list of contents.

The title of a book is even more important than that of an article. In this case you are not just persuading someone to read an item in a periodical to which they most likely already have access, but to spend some hard-earned money on something which (one hopes) will remain in use and be referred to long after many journals are gathering dust on library shelves. Additionally your title should give an indication of the scope of the book; and this will also help you to fix in your own mind the parameters within which you will be writing. One of the most succinct titles for a text book which I have come across also identified one of the best-written, comprehensive and authoritative works on its subject. *Blood Groups in Man* conveyed in only four monosyllables the content and scope of the book in a way which was an indexer's dream. With your title (at least provisionally) fixed you will have an outline target to aim at.

Within the boundaries of your title you can draft a list of contents. Start by listing some provisional chapter titles and then, over a period of several days, sort these into the most logical order and add a number of sub-headings for each chapter. When you feel that you have covered all the necessary topics you will probably be itching to start writing. It is almost certain that you will wish to revise and add to your list of sub-headings, and even of chapter titles, but don't let this worry you. Your draft contents list is like an essay plan; it is a guide, not a directive.

Unlike the writing of science articles and research or project reports a book is not something to be written as your information gathering proceeds. Boswell, in his *Life* of Johnson, reported the great man as noting that 'a man will turn over half a library to make one book'[3], and you will need to do a lot of literary research. You should have all your reference material, notes, etc. to hand before you begin or you will find that the essential 'flow' of writing will be constantly interrupted by excursions to obtain information for which you should have foreseen the need. The actual writing of a book is hard enough work, without making it more difficult by lack of advance planning.

Getting down to the writing

The business of putting pen to paper is always demanding. You will need to discover for yourself the conditions of time, space and solitude which best aid your muse—no two writers work in the same way. Whether you work best with endless cups of coffee or packets of cigarettes, and in silence or to the strains of Mozart or of Heavy Metal is something to be discovered by trial and error. Do work to a schedule, however. It is too easy to sit down with a sheet of blank paper and a blank mind to match it. Have clear ideas before you start each session both of what you are going to tackle this time and also of roughly how long you hope to devote to the session. While it is nice to find yourself carried away, with thoughts and words flowing freely so that you could keep going all night, it is also easy to overlook your own weariness—especially if you have had a hard day's work beforehand—and find later that you have been writing complete rubbish and must scrap the lot and start again.

As with essay writing, you should try to write in long continuous periods. Whether you find a pen or a typewriter best is something only you can decide, but never use pencil—it is far too hard to read when your script is eventually typed. It is quite likely that you will need to write at least two drafts of everything (maybe more) before you are content with what you have written. As with other written work, having carefully numbered all the pages put everything into a drawer and forget it for a few days. Indeed, at the first draft stage it is probably better not to worry at all about reading over and correcting what you have written until after this 'incubation period'. When you have tidied up each section or chapter into a reasonable form, and ensured that all the pages are numbered into a single sequence, put it away again until the whole book is finished—by when you will probably be sick of the whole thing anyway and in need of a holiday.

Only after another 'incubation period' should you get out your piles of paper and read through the whole work. At this stage you will probably decide that the 'flow' of ideas and of words needs re-working. Chapters do not follow-on logically from one another; some things have been repeated while others seem to have been forgotten and left out entirely; figures and tables are not in the most appropriate places and not numbered in correct sequence; you have been (unintentionally, of course) ambiguous in places and—worst of all—the whole thing makes dreadfully dull reading. This is the time to face up to the hardest and least-pleasant part of writing: the self-criticism necessary to edit your own work. You will have to be utterly ruthless, slashing out passages over which you sweated blood, consciously introducing a lighter and less pedantic style of prose, and literally taking scissors and sticky tape to re-arrange what is left into a more coherent form.

Most text books are amongst the least 'readable' of prose being turgid, pompous and formal even to the point of causing acute embarrassment to the reader. Loosen up! You cannot always write formal science as though you were dashing off a letter to a youngster, but think how you would put things into words if you were conducting a seminar for a small group of intelligent but inexperienced students. Try to use the same sort of words and phrases (but excluding the traditional 'er', 'ar', 'um' and 'you know?').

It takes some time to develop an easy-flowing manner of prose writing—especially in science—but you never will cultivate the knack if you don't unbend a little and try. Avoid slang and jargon but do aim for a relaxed style. Even in the most authoritative text book, dealing exhaustively with a complicated subject, one may find light and entertaining prose which will stick in the mind long after more prosaic and stilted phraseology has been forgotten. I have already referred to Race and Sanger's classic *Blood Groups in Man*. In the second edition, writing of 'Methods used in blood grouping' the authors commented that 'A friendly but silent atmosphere is essential. Given silence, there is still danger of the mind wandering; this it must not be allowed to do, however routine the tests may be, however primrose the alternative paths.'[4] You could do a lot worse than study this particular book as an example of how 'readable' a serious text book can be. If others can write that way, so can you.

Notes and references

Although the system of reference citations used in science papers, theses, etc. is used in books also, the lists of references may appear in any one of several ways. In some books, especially if there are many references, each chapter will have its separate numbered sequence, each starting at 1, and the list will appear at the end of the chapter. Alternatively a section of references will appear at the end of the book, consisting of separately numbered lists for all the chapters (as in this book). A third scheme—only feasible when there are not too many references in the book—is a single list with a single numbering running throughout the book. Although attractive in principle this system is most inconvenient for the author as chapters will certainly be added, assimilated into other chapters and be placed in a different order as work proceeds, causing chaos to his reference file and almost unavoidably introducing errors and inadvertent omissions. It is also a nuisance for the reader who wishes to trace a citation from the reference itself. A further system, now little used in science writing, includes the references as footnotes to the pages where the citations occur. The problems with this system are that it is inconvenient when a particular reference is cited repeatedly and—above all—it is expensive in typesetters' time, and therefore adds unnecessarily to the costs of production.

In books, even more than in articles, it may be desirable to add notes to amplify the text where you don't wish to break the train of thought or distract the reader's attention down a side turning. You may need to differentiate in your own mind between references to sources of information and notes which expand on the text. If the latter are very few in number you might get away with including them as footnotes while listing the references elsewhere. For the most part, however, you should try to be consistent in your treatment of these supplementary items throughout the book. If you do use footnotes at all it is customary to identify them with a series of asterisks, 'daggers' (†), 'double daggers' (‡), etc. and to use raised numbers only for references which are collected together elsewhere.

As with all science writing, remember that a reference which does not refer is useless. Be scrupulously careful in checking, double checking and then checking again both the accuracy of the bibliographical details of each reference and that the numbered references in your list agree with the numbers of the citations in your text.

The preliminary and end pages

A book has a number of elements additional to those of an article and if you would impress a publisher with your ability to produce a carefully thought out typescript you will not ignore these. The preliminary pages—known in publishers' jargon as 'prelims'—are the necessary additional pages which come before the main text while the end pages (not surprisingly) come at the end. Unless you have had occasion to study the precise layout of the prelims you may not have realized how, in all books, they follow a format as structured as that of any research paper. Study some books at random and you will see what I mean. You should endeavour to meet the requirements of that format as nearly as possible in your typescript. The prelims are usually made up as follows:

Half-title page. This is the first printed page of the book and carries only the title on an otherwise blank sheet.

Half-title verso. Strictly speaking *verso* means 'to turn over'. In publishers' jargon it is the left-hand page behind a main entry. Printing convention is that all new items of text begin on a right-hand page and right-hand pages carry odd page numbers beginning at 1. Verso pages, therefore, are always even-numbered. The half-title verso is usually left blank but it sometimes carries a list of other books by the same author or in the same series.

Title page. This is the main title and apart from the title of the book it carries the author's name, sometimes with his or her qualifications and/or affiliations, and the publisher's name and 'device' (sometimes wrongly referred to as his colophon).

Title verso. This carries a copyright notice (the wording of which may be prescribed by the individual publisher but which is usually standard); an International Standard Book Number (ISBN) which is provided by the publisher; and a statement of where and by whom, and for whom, the book was printed. This information is also supplied by the publisher.

Dedication. If there is a dedication or quotation then it usually appears on the next right-hand page (known as a recto: the opposite of verso). In this case the following verso will remain blank. To fit in with printing layout, however, a dedication will sometimes be fitted in to the Title verso.

Contents list. Basically this is a list of chapter headings but if the chapters are divided into sub-sections you should also list these for each chapter. The list should simply be headed 'Contents', and not with the word 'list'. Obviously you cannot fill in final page numbers on a typescript but it will help you (and your publisher) to find your way around the typescript if you write in pencil the relevant numbers of the typed pages.

List of Illustrations. If you have more than a very few illustrations you should list them here, giving the text of the relevant legends. Again, you could pencil-in the approximate position in which they should appear by giving the page numbers of the typescript.

Preface or Foreward. Not all books need either of these items but if one is called for, this is where to put it. A preface has been described as 'the introductory address of the author to the reader, in which he explains the purpose and scope of the book'. A Foreword, on the other hand, is written by someone other than the author—often someone eminent in the discipline concerned—to introduce the author and/or his book. Today, most people prefer to judge a book on its merits without a 'plug' from someone else, and if a book needs to be justified, or its purpose explained other than in the 'Introduction' chapter, then it probably has no market and few publishers will take it on.

The end pages of a book are simpler than the prelims. Immediately after the text should come the appendixes. Generally you should only include an appendix to amplify matter contained in the text when it would be inappropriate for it to be part of the text. In science the sort of things which are best given in this form are lists of definitions, mathematical and conversion tables, lists of normal physiological values and chemical constants, etc. These are followed by notes and references and/or a bibliography (a fully-detailed list of books for further reading on the subject, prepared in the same style as your reference citations). Any Acknowledgements also appear in the end pages just before the final item, the index.

An index is essential to any factual book. Sometimes a publisher will prepare the Index but if you have to do it yourself you will find that it is quite a chore. You will need a large number of cards. Going through the text of the book you write onto each card a word which seems to you one which readers may wish to refer to. Generally these will be nouns or verbs. Let your own experience as a reader and as a student guide you as to what is and what is not worth including. When the cards are all complete you should put them in alphabetical order and wait until you have page proofs which carry the final page numbers, which you must then add to the cards. Some words will appear only once and others several times. If in one place an item covers two or more pages, all dealing with the same subject indexed under one word, then indicate this thus:

Freezing, of red cells, 282–3, 384
Genes, 5–6, 10, 12
Genetics, 5–15

Whether you will need to have your index re-typed as a list is a matter for you to discuss with your publisher: some publishers are prepared to work from a card index, others are not.

In your final Index you should add *n* to any page number where the item appears in a footnote and a wavy underlining to any page number which refers to an illustration to indicate that the number should be printed in bold type. A note explaining this convention should be included at the beginning of the index.

Your publisher

So far we have been assuming that you are writing your book on your own initiative and that you will in due course submit it to a publisher much as you would submit an original paper to a journal for possible publication. This is a considerable act of faith if you are not already known to any publisher as an experienced writer, for the investment of time and effort which goes into writing even a short book is considerable.

Of course, it does not always happen like that. Some authors of text books, if they are particularly confident of their market (and their own writing ability) send a copy of their proposed outline and a sample chapter to a number of publishers to assess their reactions. In effect the author is putting his (as yet unwritten) book out to tender and soliciting offers—but there are two words of warning: honesty and caution. If you offer your book to more than one publisher at a time it is only honest to tell them what you are doing: and do make sure you have at least some of the rest of the book written. One swallow does not make a summer and if a publisher is interested he will almost certainly wish to see the rest of the book before committing himself. 'Commissions on part of the material only are rarely given to untried authors.'[5] You may be lucky, however.

Sometimes, if you have achieved a reputation either as an original worker or as an outstanding teacher you may find that a publisher approaches you. You may even start your book-writing career by being asked by a professional colleague to contribute a chapter or two to a multi-author book which he is compiling or editing.

If you are starting 'cold', however, and have produced a typescript of your own volition you will need to find a publisher willing to take a risk with it. Selecting the firm of publishers that you would like to accept your book may not be too difficult: convincing them of the suitability of you and your book could be more difficult. As a practising scientist you have one advantage over most other first-time authors. It is your professional duty to be familiar with the literature of your subject and you are almost bound to be aware of which publishers are responsible for the majority of valuable books in your own discipline. If you have never noticed which firms publish the books which you regularly read and consult professionally it will not take you very long to check this and produce a short list of two or three which seem to you to be the most suitable for your own book.

Before you hand your manuscript to a typist you should take great care to see that it conforms to the 'house style' of the publisher to whom you propose sending it. 'House style' is simply the preferred style for details such as the use of capital letters, punctuation (especially of abbreviations), style of headings and subheadings used, method of quoting references, etc. Many publishers issue a printed or duplicated sheet of guidance for use by authors and their typists; others refer to printed books of style which are available (usually to special order) from booksellers. In any case one should take three or four recent publications of the publisher and study these carefully, using them as a guide to what is or is not likely to be acceptable to that publisher. Whatever else, you do, however, be consistent within your typescript if you wish to be taken seriously.

Your instructions to your typist should then be much the same as those suggested in Chapter 6. The text should be typed on A4 size paper with double spacing and wide margins, and the approximate position of illustrative material indicated in the margin. Tables and legends to figures should be kept separate from the body of the text and, most importantly, all the pages should be numbered sequentially—preferably in the top right-hand corner—so that if they become separated neither the publisher's reader nor you will be left trying to work out which goes where.

Publishers are always delighted to receive perfect typescripts, free of all errors and corrections, but if you have to correct typing errors or make deletions this is not total disaster. As long as any corrections are made neatly and legibly your typescript will not have to be (expensively) re-done. If a page has more than one or two simple corrections on it, however, you may be better going to the trouble of having it re-typed rather than risk making a bad impression.

Submission—and after

When your manuscript has been typed do not be tempted to bind it into a ring binder—or any other form of binding—for the purpose of submission. Make sure you have at least one extra copy (xerographic rather than carbon) which you can retain, and be prepared to provide the publisher with a further copy if he subsequently asks for it. Fasten the pages of the typescript together in chapters, each retained by no more than a paper clip or a treasury tag, and put the whole lot into the box which held your blank typing paper. Publishers' readers study typescripts in all kinds of places and a train (or a bed) is not conducive to easy handling of anything other than loose sheets. You will possibly find it more convenient—and safer—to make xerographic copies of your illustrations and clip these to the typescript, placing the originals in an envelope separate from the typed pages.

Your covering letter is particularly important. Publishers are besieged by would-be writers and one of the first things they wish to know is whether you can write. Don't telephone, don't call personally, and don't scribble a brief note saying you want to submit the enclosed book for publication. Compose your letter as carefully as you composed the book itself, check and revise it as carefully and have it typed as carefully. A good opening impression is well worth the effort of making it.

After an initial acknowledgement of receipt of your typescript be prepared to wait for two or three months before you hear any more. It takes the very few people who perform this task a long time to give sound and critical judgements on the large number of works which flow into a successful publisher's office each day. When you do hear the response will be much on the lines of those given by an editor to authors of science papers: accepted; accepted with modifications; or rejected. Your response (after the initial joy, modified rapture or shattering disappointment) should also be on the same lines; celebration; thankful agreement to undertake the necessary extra work; or a radical re-appraisal and re-writing before trying again elsewhere.

If your book is accepted you will be offered a contract and in general these are fairly standard amongst reputable publishers. Normally you will be offered a royalty payment on each copy of the book sold—usually starting at 10% and possibly rising to 12½% or 15% depending upon the number of copies actually sold. Do be careful of what is being offered, however. For the author, royalties are best based upon the published price of the book but many publishers base them upon net sales receipts. This simplifies the publisher's accounting but as he sells books at discounts to suppliers, and may also supply copies on a sale or return basis, the author comes less well out of the arrangement. A (not uncommon) average discount of 40% to suppliers by the publisher would itself require an author's royalty of 16.6% on net sales to be equivalent to a royalty of 10% on the published price. If you can reasonably expect a number of North American sales of your book examine carefully the royalty arrangements for these, which should bring the author not less than 10% of the net receipts. A reputable publisher should agree to pay you an advance against royalties and this might reasonably be 65% of what it is estimated you should receive from the sale of the first printing.

The contract will also provide that you will receive half-a-dozen complimentary copies of the finished book as well as having the opportunity to buy further copies at a discount (usually 33⅓% or more) off the normal price, providing you agree not to re-sell them. You will also probably be asked for any suggestions you may have for marketing and advertising the book amongst your professional colleagues, and requested to provide a short 'blurb' about the book which will be printed on the jacket.

Do be prepared to argue with your publisher about any terms in the contract which you may feel to be unfair. The publisher may indeed be offering exactly what he feels to be right in your case, or he may be like a professional bidder at an auction—knowing what his ultimate limit is but starting the bidding at a lower level. Remain friendly in your approach but be firm if you feel you are being offered a raw deal. In 1981 the Society of Authors and the Writers' Guild drafted a 'Minimum Terms Agreement'[6] which is eminently fair to authors (and which some publishers are prepared to offer to authors who are members of either authors' organization). You could do worse than take this as a model and try to negotiate a contract for yourself based upon it.

After your finalized version has been received by the publisher it will probably be between six and twenty months before the book is actually published. Don't be dismayed at such a delay; it is due to many complex factors in both printing and publishing industries and is quite normal. Only very exceptionally does a book appear in less time than this.

Do not expect to make a fortune from having a book published: you may not even cover your literary research and secretarial costs. You will probably not achieve instant fame, either. But there is a definite sense of satisfaction, as well as pride, in seeing your project fulfilled. It is rather like having children. Your offspring may succeed or fail; it may survive for a long life or die an early death; you may be proud of it or come to refer to it as little as possible. Whatever the outcome, you will have created something unique: and parenthood can be a mighty fine feeling.

Chapter 9
What about word processors?

Word: a basic unit of computer memory
Computer: a device which can accept and supply information

D. Godfrey and D. Parkhill (Eds) in *Gutenberg Two*, Toronto: Porcepic, 1979, pp 221, 199.

Word processing is using a computer to manipulate text with the aim of perfecting documents

A.J. Asbury. Word Processing. *Brit. med. J.* 1983; **287**, 44.

Do we have a problem?

'People have been writing on paper or papyrus for four or five thousand years—long enough, anyway, to get into the habit.'[1] When Shakespeare, Milton and the Brontës produced their masterpieces they each took a pen, dipped it in ink and wrote upon paper. Not until 1875 did the American humorist Samuel Langhorne, writing under the pseudonym 'Mark Twain', become the first author to submit to a publisher a manuscript which had been written on a typewriter (one of the first 400 mass-produced by the Remington Small Arms Co. and which had cost him $125.00). In the succeeding hundred years the terms 'manuscript' and 'typescript' have become effectively synonymous in the publishing business and until recently the typewriter was the only writing instrument used to produce works for publication.

The development of the micro-computer has changed all that, although even in 1984 one software manufacturer is saying in a sales promotion leaflet that 'ninety percent of the business executives in this country suffer anxiety, insecurity and sweaty palms at the mere thought of using a personal computer'. It is probably true to say, indeed, that the western world is divided by a generation gap into two groups of people: those who understand computers and those to whom they represent 'a riddle wrapped in a mystery inside an enigma'[2]. Nevertheless it has been recognized that 'a new generation of authors is growing up with computers (and) a keyboard may become more familiar than pencil to many of them'[3].

Of all people, scientists are to the fore of those who have 'taken' to computers. For storage, retrieval and manipulation of the raw data of their investigations, as well as for the more mundane business of stock-control of consumable stores, the computer has proved an invaluable tool. It is little wonder then that the ready-availability of micro-computers with word processing packages has led many to use them for the production of research papers and text books. Unfortunately the first law of computer usage—garbage in = garbage out—continues to apply and professional writers have often had more success in using word processors as a writing tool than have professional computer users.

So what are the constraining factors in achieving literary success with a word processor? Essentially there are three: (i) the ability to write lucidly—whatever the medium; (ii) a thorough appreciation of what the word processor can and cannot do–and how to do it; (iii) an understanding of the hardware requirements—which may differ from those for data processing.

On the assumption that any scientist contemplating writing on a word processor rather than with a typewriter has access to a computer in his scientific work, and therefore understands the jargon and the basic principles of computer usage, we shall take these as understood. The ability to write lucidly is what this book is all about, so that also we shall take as read at this stage and go straight to considering just what a word processor can (and cannot) do, and how to make the best use of it.

What can—and can't—a word processor do?

The basic advantage of writing on a word processor is that electronic capture of the work when you first depress the keys offers great cost savings in paper and time for both you as the author and your publisher. If instead of a typewriter the keyboard is on a micro-computer the text which is keyed-in can be edited on the screen of the visual display unit (VDU), and this is both a quicker and a more legible process than using white ink, rubbers, scissors and paste, red pens and crossing-out on paper, crumpled sheets of which overflow from most writers' waste-paper baskets. Once the immediately-obvious errors have been corrected the text can be transferred onto a floppy disc where it can be stored until needed: and when subsequently called up again words, phrases, sentences and whole paragraphs can be deleted, added, altered or moved about without the need to re-type the whole thing, and the revised version then stored again. When the author is finally satisfied (or at any time previously) he can feed the contents of the disc to a printer and obtain hard copy of what he has written. This can have all kinds of effects not easily obtained with a typewriter such as bold-face type, automatically centred headings, justified (i.e. straight) right hand margins, automatic page numbering and automatically indented blocks of text.

Now to do all this it is necessary that the micro-computer has the necesary peripherals—a VDU, dual disc drive and a printer—and that a suitable software package is used. All of these we shall consider further, later in this chapter.

Given these essentials the potential of word processing becomes almost unlimited. At the simplest level the author can produce a number of hard copies of what he has written for submission to a journal or a book publisher in the conventional manner, just as though it had been produced on a typewriter. Becoming more sophisticated, the floppy disc itself (or a copy of it) can be submitted: this allows the editor (or publisher) to edit it electronically on his own computer. Peer review can be conducted in one of three ways: (i) hard copy can be supplied to the referees in the conventional manner; (ii) the floppy disk (or another copy) can be supplied for display on the referee's own computer; or (iii) using a MODEM (MODulator-DEModulator) the digital information on the disc can be transmitted by telephone directly to the referee's computer. Once accepted, an article or a book can be copy-edited electronically to provide a layout suitable for printing with headings and sub-headings in different type sizes and styles and finally the edited disc used (possibly *via* another MODEM) to drive a digital phototypesetting machine—which may have available over 1000 different typestyles in a range of sizes from 4 to 72 point (i.e. 1/18″ to 1″)—to produce the final printed copy, 'untouched by human hand'.

Now if all of this sounds like George Orwell's *Nineteen Eighty Four* there are two points to be made. Firstly it is all happening, here and now in 1984. Secondly it is so fraught with practical difficulties—scarcely any of them electronic—that few people are yet able to take full advantage of what technology now offers. This will change, however. At an international conference of scientific editors held in 1980 it was said of word processors that they 'present advantages and disadvantages, and whilst possibly at the moment the disadvantages outweigh the advantages, in the foreseeable future the reverse will be the case'[4]. (It was also noted that 'already in about one in ten word processing installations photo-typesetting is completely tied into the word processing operation'[4].) So what are the advantages and disadvantages?

Apart from substantial savings in keyboard time (and paper) when it is no longer necessary to type an article or a book several times in order to get it right, the advantages include for the publisher greater ease of communicating with the abstracting services, for the publisher or the editor considerable simplification of indexing, and for the author no need to rely on a typist to translate his (often illegible) handwriting into a neat typescript laid out exactly as he wants it.

The disadvantages are more numerous. Firstly, of course, one must have access to suitable hardware. If you have not and are contemplating buying your own be warned—it is expensive (see the next section). If you believe you have, you may be disappointed. Not all micro-computers are suitable and most printers used in laboratories for data processing are unsuitable. In any case, will your hardware (or software) be compatible with that used by your target journal or book publisher? A floppy disc from your Apple™ micro-computer may not be much use to the editor of your target journal if he uses a Sirius™ or an IBM™. Of course, if you intend using your word processor only to produce hard copy for submission this difficulty will not apply. Even in this case, however, you will probably be tied to your keyboard when writing your paper (or book). Unlike pen and paper a micro-computer (even a small one) cannot readily be used for literary composition in your bed, bath or bus. However, portable word processors are now available which are suitable for composition but need conventional peripherals for displaying blocks of text and for printing hard copy.

If it lives in your laboratory that is for most people probably the only place where you will be able to use your micro-computer to commit your deathless prose to posterity, to edit and to print it. (Your journal editor may feel the same constraints if you send him a floppy disc. Many editors of science journals are also practising scientists and do their editorial work almost anywhere except in their own laboratory.)

What few scientists who are keen to do their writing on a word processor appreciate is that the use of computers pushes the publishing process back towards the author. Much of the preparation of text for printing which was previously the province of the printer is pushed back upon the editor and copy editor who may have to enter generic codes onto the floppy disc to drive the phototypesetter to produce the desired layout. In turn, much of the work done at present by the editor—correction of typing, spelling and grammatical errors, standardizing nomenclature, etc.—is pushed back to the author. In effect, authors will have to produce perfect 'camera ready' script themselves: and 'one can only hope that their literary abilities keep pace with their technical competence'[5].

Finally, a substantial disadvantage of word processors is that while they can produce text satisfactorily they are very restricted in their ability to produce graphics. If you wish to illustrate your work you will find that no more than the most rudimentary forms of illustration are possible. Tables are reasonably simple; graphs and histograms can be incorporated into discs carrying word processor text only with difficulty and at great expenditure of time and expertise; and half-tone photographs are simply not a possibility. Of course, if you are using your computer merely to produce hard copy for conventional subsequent handling this is not a problem. If you are aiming at submitting your work on disc, however, then you are going to be very restricted indeed in your choice of graphics.

Many authors may prefer to hasten slowly in this field and follow the advice of a Canadian publisher who said that 'in my own organization we follow the practice of Royal Consorts and keep two or more paces behind the leaders. If they trip, we hope to profit from their errors'[6].

Getting the right hardware and software

If you are seriously considering doing your writing on a word processor then there are three likely possibilities: (i) you have access to a business micro-computer with the necessary peripherals and a word processing package; of (ii) you own a personal micro-computer and contemplate adding further peripherals to it and/or buying some more software; or (iii) you are considering buying a personal computer for the first time and see word processing as a good excuse for taking the plunge. In each of these cases you may find that the minimum requirements are less straightforward—and more expensive—than you think.

Personal micro-computers have become almost as plentiful as pocket calculators and while in 1984 one can buy a computer and a television screen for use as a VDU for under £100, and even set up a workable word processor for a couple of hundred pounds, a satisfactory outfit capable of producing the kind of results which will be acceptable to an editor or a book publisher, and will be worth the outlay and the effort, is unlikely to leave much change out of £1200; while a quite modest professional business set-up will run to £2000 or more. If you are going to shop around for the best buy then you will need to know the minimum specifications for each item, so let us consider them in turn.

The computer

As you will be using this in lieu of a typewriter a high-quality typewriter-style keyboard with the standard QWERTY layout is essential, and obviously there must be the necessary expansion sockets and interfaces to support the peripherals. Less obviously, to store your text while editing your central processing unit (CPU) must have an absolute minimum of 32K RAM memory—64K is infinitely better. As you will quickly discover, many of the more readily available home computers do not offer this although all professional quality micros certainly will. You must also have a cursor that can be moved all over the screen.

As one of the objects of the exercise is to produce text on A4 size paper, and to be able to edit the text on screen, the computer needs to be able to support a screen with at least 16 lines, each 80 columns wide (this being the number of characters and spaces in an average line of type across the width of an A4 sheet). Word processing software for home computers often has to adopt devious devices to fit in with a more restricted screen width of 23 to 32 columns (and rarely more than 64), which means that you can not view the whole width of a line of text at once. Often the cursor remains stationary while the text passes by as you enter it: this not only makes editing tedious, it also impedes the flow of your thought if you cannot scan the whole of what you have just written from time to time as you are writing. You should also ensure that your micro can display lower case characters as well as capitals on the screen: not all home computers do so and this could be a great inconvenience when using much scientific terminology.

The screen

This may be variously described as a visual display unit (VDU), cathode ray tube (CRT), or simply as a monitor. Most home computers use a television set as a monitor but for word processing this will not suffice. Monitors are available in various sizes but a 12-inch (30 cm) screen is most commonly used. Colour is unnecessary and in order to cause the least eye-strain with constant scanning of text a monochrome screen which displays green or amber characters on a green, brown, grey or black background is generally the most satisfactory.

The reason why television sets do not make good monitors for word processing is that they do not give sufficiently high definition. Text and graphics are blurred and not sufficiently easy to read, and the picture tends to 'bend' around the edges of the screen. Additionally there is an irritating flickering caused by the horizontal lines which make up the picture moving down the screen at a different rate to that scanned by the CPU. If only to save yourself perpetual eye-strain and headaches you should use the highest quality dedicated monitor you can acquire (or afford) for word processing.

Text and program storage

For word processing you will require two sources of storage outside your CPU—one to hold the word processing program and one for you to transfer your writing to, in blocks as you compose it, and from which you can retrieve it for further editing and/or printing. In practice there are three methods of external storage possible, magnetic tape cassettes, floppy tapes and floppy discs. Although the cheaper home computer word processing systems utilize cassettes these are quite unsuitable for creative writing such as the composition of scientific papers or of books. Apart from their relatively limited storage capacity cassettes are typically some 60 times slower than discs for accessing the material stored on them: and the main reasons for using a word processor rather than a typewriter, correcting fluid and reams of paper are speed and ease of editing. Also, the better word processing software is all disc based. Floppy tape is faster than cassettes and cheaper than discs, but available on very few systems as yet.

Disc drives tend to be amongst the more expensive peripherals, as well as the hardest to assess for storage capacity. As with so much in computing, the only answer to 'how much storage?' can only be 'as much as you need'. For a compromise between reasonable storage capacity and reasonable cost you should aim to use 5¼″ double-sided (and preferably double-density) floppy discs with dual drive units giving you a total storage capacity of 400 K or more. Of course, if you have access to something better than this it is all to the good.

The printer

For all practical purposes you will only be concerned with impact printers and these are of two main types, dot matrix and solid-type. In each case there is usually no keyboard and the input is solely from the CPU. Virtually all printers have a data buffer—a temporary store for data coming from the CPU faster than the printer can print it—and this may be as much as 16 K.

Most printers used with laboratory-based and home microcomputers are of the dot matrix type, forming characters composed of a series of dots which may be five dots wide by seven dots high to nine dots by fourteen, depending on the printer. Not all such printers have lower case type with true descenders (i.e. the tails of the letters g, j, p, q and y descending below the line) and in this case the printed text is often not easily readable and will most likely not be acceptable.

The advantages of dot matrix printers are that they are cheaper, faster (from 10 to 500 characters per second) and less noisy than the alternative daisywheel and thimble printers: however, due to poor quality of the type they are not all suitable for producing hard copy of letter quality for submission to journal editors and book publishers, some of whom will not even consider such submissions unless your printer does produce text of letter quality[7].

The alternative daisywheel and thimble printers are more often found in office-based systems as they produce type suitable for business letters and which may be distinguishable from that produced by a good electric typewriter. Unfortunately, apart from being expensive they are also noisy: it is usually impossible to carry on a telephone conversation within 20 ft (6 m) of one when it is operating. Printing speed, at 8 to 80 characters per second, is slower than with dot matrix printers but is still quite fast.

A more recent innovation has been the dual-purpose electronic typewriter/printer, which is relatively inexpensive and is more versatile than a dedicated printer being capable of a keyboard input as well as that from the CPU. They are relatively slow and noisy but produce an excellent print quality and are useful if the interface matches that built into (or available for) your CPU.

A possible solution if your laboratory micro has a dot matrix printer which is not of letter quality is to look into the departmental office. An IBM 'golfball' typewriter can make an excellent printer if you have one accessible, and suitable interfaces are readily available.

Finally a word about paper. Computer paper is usually continuous and fan-folded: only use paper folded to produce a vertical format and if you are intending to submit your literary work as hard copy only use plain white paper of 70–90 g/m^2 weight. Submissions in a horizontal format or on lined or 'music-ruled' paper may well be returned to you un-read, and lighter-weight paper may crumple and tear easily. And do not fail to separate the individual sheets of typescript and remove any perforated edge-strips before sending them off for consideration.

The program

To use a computer for word processing you need a special program and these are available mostly on floppy discs, although there are some programs on plug-in ROM chips, at least one on floppy tape and (despite the unsuitability of the hardware) even some on cassette tape. Which program you use will largely depend on what is available for your computer. Most micro-computers use the CP/M™ (Control Program/Microcomputers) control system based on Z80/8080 micro-chips and you will find that nearly all the better word processing packages (e.g. Wordstar™) are designed to work with such systems. However, a number of popular micro-computers (e.g. Commodore™, Atari™, Texas™, Data General™ and most Hewlett Packard™) use other chips and the choice of word processing software for these may be limited. The Apple™ and BBC™ are special cases: although operating on a 6502 chip a plug-in Z80 chip is available which enable them to use CP/M-based software.

Don't get too worried about the finer points of electronics and micro-chip technology at this point, however. Seek advice from a reputable software supplier and obtain the most powerful (i.e. flexible) word processing package that your budget will run to and which is compatible with your hardware. But beware! Software is expensive. Do not buy it 'blind' but have your dealer demonstrate it. Check what happens if you deliberately press the wrong key at a crucial point, or enter a date like 31 February: and see if you can understand the users' manual provided. A good supplier will provide a warranty and after-sales service, including a telephone 'hotline' for emergency queries. If you have found such a supplier then put yourself in his hands: if not, then keep shopping around until you do.

How to set about it

As this book is about writing, a detailed guide to operating a word processor would no more be in place than a chapter devoted to typing technique. I must assume that if you have got this far you known how to operate a micro-computer. However, you will probably not be familiar with your word processing program and I can only entreat you to spend some time finding out how to operate it and determining what exactly it can do. If you have a powerful program you may never fully understand and utilize (or need to) all the facilities it offers, but do take time to master as many as them as you can. A good program will be menu-driven so that careful study of the screen and of the manual (and maybe some other guides to the system) should soon make all clear, providing that you know what is possible.

Having booted-up your computer, initialized your data disc and opened a new file you are ready to start writing. Seated at the keyboard you can at first treat it exactly as though it were a typewriter. As you type in your thoughts they will appear on the screen, but with the difference from typescript that you will probably have automatic formatting. The facility known as 'word wrap; means that you don't have to push the equivalent of a carriage return at the end of each line: when you reach the end of one line the program will automatically move on to the next. Depending on the sophistication of the software it will also probably produce a justified right margin and may either split and hyphenate long words at the end of a line or remove them completely to the next line, filling the first line out with spaces between the other words.

The great advantage of a word processor is that you can edit on screen so that typing mistakes ('typos'), spelling errors and second thoughts about choice of words, punctuation, etc., can be dealt with as they occur, leaving a neat and tidy text.

Having put your thoughts together you will naturally want to preserve them. So far they are no more than electrical impulses held tenuously together by a micro-chip and a few printed circuit boards. If the power is turned off (or fails) or the program crashes, your words will be lost and you will have to start all over again. Most programs will transfer material automatically to the data disc once you have typed in as many words as will fill the RAM storage capacity of your CPU. As further text is typed in so the earlier material is automatically transferred to disc, where it is safe. However, if something goes wrong you will lose everything which is still in the CPU memory: and that can be a lot of writing—may be 1000 words or more. The solution is to transfer blocks of material to disc at regular intervals; a procedure which requires only a simple key operation. A good program will indicate the end of each page (usually by means of a broken line across the screen) and this may be a convenient point to make your transfer: or you could do it after each paragraph.

As with ordinary writing techniques you should put your composition away for a time before going through it carefully for second thoughts. In this case, however, you can either put your storage disc in a drawer and subsequently review it on screen or you can print out hard copy for conventional review and correction. If you do it this way you will have the advantage of being able to edit the disc in line with your hard copy corrections but without the need to re-type the whole thing, and you can print the revised version with no more hassle than depressing a few keys on your CPU keyboard.

If you intend using your word processor to produce hard copy for conventional submission to an editor or a publisher there is little more to do beyond any further refinement of your second, third or subsequent thoughts. If you intend submitting your work on disc then there may be a good deal of work left in formatting your text into a layout more suitable for printing. The extent of this use of special print features will vary with the arrangements you may make with your editor or publisher. A really good combination of program and printer is capable of producing hard copy of professional print quality for use as 'camera-ready' text, while a carefully produced disc can be used to drive a phototypesetter. The quality of the disc in either case will directly reflect your skill in fully utilizing the facilities of your software, so do not aim to submit anything on disc unless (i) you have fully mastered the program you are using and (ii) you are sure that your target journal or publisher is prepared to accept such submissions.

And finally

There is no doubt that word processors have a big future in the writing and publishing business. They are already revolutionizing journalism and many professional writers in other fields are using them. Increasingly, scientists will find this the most convenient way to produce research reports, text books, etc. However, do not forget that a word processor is no more than an instrument for setting down thoughts in a form which can be preserved, just like a pen or a typewriter. The thoughts must be there in the first place, as must the writing skills which this book is all about. For any writer the first essential is that he has something to say; next comes his ability to say it lucidly and coherently; only then comes mastery of the technology for recording it.

Appendix
British proof correction marks
(BS 5261 :1976)

Instruction to printer	Textual mark	Marginal mark
Insert new matter	λ	New matter followed by concluding stroke (/)
Delete	Stroke through character(s) to be deleted	∂/
Delete and close up	∂ through character(s) to be deleted	∂
Close up	⌢⌣ between character(s) or spaces to be closed up	⌢⌣
Insert as superscript	λ	λ
Insert space	# or > if between lines	#
Invert block or character	Encircle block or character to be changed	9
Align text matter		||
Re-arrange characters or words in correct order	⊔⊓ between characters or words to be transposed	trs./
Change to italic type	—— beneath character(s) to be changed	ital
Change to bold type	~~~ beneath character(s) to be changed	bold
Change to capital(s)	≡ beneath character(s) to be changed	cap(s)
Change to lower case (small letters)	Encircle character(s) to be changed	l.c.

Example of proof correction

ORIGINAL TEXT WITH CORRECTIONS

Leucocyte-poor blood

The fficiency of various proceses for removing Leucocytes can vary widely. Complete removal is virtually unattainable and for this reason the term 'leucocyte-poor' bloood (or red cells) is preferred to 'leucocyte-free'. generally there have been five methods of preparation used: centrifugation inverted, dextran sedimentation, nylon filtration, recovery of frozen cells, and saline washing. For all these methods the donation of whole blood has usually bein required to be notmore than 24 h old, as with older blood the clumping of leucocytes and platelets can make them difficult to remove—although this view point has been
challenged.

Med Lab Sci (1983) 40, 105

bold/ e/ s/ l.c./ trs./ cap/ trs./ #/ ital./ bold/

AMENDED TEXT AS PRINTED

Leucocyte-poor blood

The efficiency of various processes for removing leucocytes can vary widely. Complete removal is virtually unattainable and for this reason the term 'leucocyte-poor' blood (or red cells) is preferred to 'leucocyte free'. Generally there have been five methods of preparation used: inverted centrifugation, dextran sedimentation, nylon filtration, recovery of frozen cells, and saline washing. For all of these methods the donation of whole blood has usually been required to be not more than 24 h old, as with older blood the clumping of leucocytes and platelets can make them difficult to remove—although this viewpoint has been challenged.

Med Lab Sci (1983) **40**, 105

An annotated bibliography

> Some books are to be tasted, others to be swallowed, and some few to be chewed and digested.
>
> Francis Bacon (1561–1626)
> *Essays. Of Studies.*

Chapter 1
Writing good English

Fowler HW, Fowler FG. *The King's English*. 3rd edn. Oxford: Oxford University Press. 1970.
First published in 1906 this is one of the classics of English usage and is a very detailed study of grammar. For the serious student only.

Gowers E. *The complete plain words*. 2nd edn, revised by Fraser B. London: HMSO. 1977. (Also in paperback. Middlesex: Penguin Books).
The modern classic on use of English. Very comprehensive and very readable; full of hilarious examples of mis-use of the language. Highly recommended.

Butt M. *Dictionary of English usage*. London: Collins (Gem series). 1970.
The most concise, comprehensive and pocketable book on English at a pocket money price. If you can only afford one book on the subject get this one.

Hawkins JM. *The Oxford minidictionary*. Oxford: Oxford University Press. 1981.
You must have a dictionary: this one has nearly all the non-technical words you will ever need and matches Butt's book (above) in price and size. A real bargain.

Onions CT (Editor). *The shorter Oxford English dictionary*. 3rd edn. (2 vols). Oxford: Oxford University Press. 1973.
If you want a big splurge on a really good and authoritative dictionary this is the one to go for.

Roget's thesaurus of English words and phrases.
(There are many editions of this work available both in hardback and paperback, including an inexpensive one in the Pan Reference Books series.)
First published in 1871 this classic work is invaluable in its modern editions as a guide to words of like meaning and of opposite meaning. If you're ever stuck for a word, you'll need an edition of *Roget's Thesaurus*.

The Oxford dictionary for writers and editors. Oxford: Oxford University Press. 1981.
An authoritative dictionary with an emphasis on words which often get mis-used or mis-spelled. Useful abbreviations, foreign words etc.

Partridge E. *Usage and abusage*. Middlesex: Penguin Books. 1973.
Subtitled 'A guide to good English' this is another dictionary of English usage which wittily attacks linguistic abusage of all kinds. Very readable and in a cheap edition.

Carey GV. *Mind the stop*. Middlesex: Penguin Books. 1977.
An excellent, short and inexpensive guide to punctuation. Every writer should have a copy.

Partridge E. *You have a point there: a guide to punctuation and its allies*. London: Routledge & Kegan Paul. 1978.
A very detailed standard work on punctuation for anyone with a serious interest in the subject.

Chapter 2
Writing essays

Lewis R. *How to write essays*. Cambridge: National Extension College. 1976.
The text-book for a correspondence course in essay writing—but it stands on its own as a structured guide to how to write essays.

Chapter 3
Writing science papers. 1

General notes on the preparation of scientific papers. London: Royal Society. 1974.
A short and very basic pamphlet which is useful for reference—if you have access to it.

Booth V. *Writing a scientific paper and speaking at scientific meetings*. 5th edn. London & Colchester: The Biochemical Society. 1981.
A concise—and rather terse—pamphlet on the lines of that of the Royal Society (above), but with a section on public speaking. Also not widely available.

Day RA. *How to write and publish a scientific paper*. 2nd edn. Philadelphia: ISI Press. 1983.
An excellent readable and witty step-by-step introduction to science writing. Unfortunately not generally available in the UK.

O'Connor M, Woodford FP. *Writing scientific papers in English*. London: Pitman Medical. 1979.
A very comprehensive guide. The style manual of the European Association of Science Editors.

Lock S. *Thorne's better medical writing*. 2nd edn. London: Pitman Medical. 1977.
Written by the editor of the *British Medical Journal* and specifically addressed to medical practitioners. Very readable.

Huth EJ. *How to write and publish papers in the medical sciences*. Philadelphia: ISI Press. 1982.
An excellent American equivalent to Lock's book (above). Very much a book for the serious and/or experienced writer. Not generally available in the UK.

CBE Style manual. 5th edn. Bethesda Md: Council of Biology Editors. 1983.
The standard American style manual. For the serious student of science writing only—but hard to get hold of in the UK.

Units, symbols and abbreviations. 3rd edn. London: Royal Society of Medicine. 1977.
An excellent short guide to the use of units, etc. Worth the effort of getting a copy for reference.

The SI for the health professions. Geneva: World Health Organisation. (Available in the UK from HMSO). 1977.
A most valuable short paperback for those uncertain about the use of SI units.

Chapter 4
Writing science papers. 2

See all the works listed for Chapter 3, also:

Bibliographical references in scientific publications. London: Ciba Foundation. 1980.
Suggestions from an ELSE-Ciba Foundation workshop. Useful for the experienced writer but not readily available. (Bound with *Camera-ready typescripts*. See Chapter 6.)

Bibliographical references. BS 1629:1950. London: British Standards Institution. 1951.
If you need the British Standard, this is it.

Reynold L, Simmonds D. *Presentation of data in science*. The Hague and London: Martinus Nijhoff. 1981.
If you want to become your own illustrator this is the text-book you need.

Flint MF. *A user's guide to copyright*. London: Butterworth. 1979.
Everything—and more—you may ever need to know about copyright. Even a layman can understand the complexities of copyright law with this guide.

Chapter 6
How to prepare . . . your paper, and deal with editors and printers

Hart's rules for compositors and readers. 37th edn. London: Oxford University Press. 1983.
Although originally a 'house manual' for the Oxford University Press this little volume is the definitive work to give to your typist if you want to get your paper in the correct format.

Camera-ready typescripts by authors and typists. London: Ciba Foundation. 1980.
If your target journal demands a camera-ready typescript may heaven help you. Both you and your typist will need this booklet. (Bound with *Bibliographical references*—see Chapter 4).

Copy-preparation and proof correction. Part 2. BS 5261:1976. London: British Standards Institution. 1976.
The definitive standard in the UK for proof correction. It's doubtful if you'll ever need the whole thing.

Butcher J. *Copy-editing*. 2nd edn. Cambridge: Cambridge University Press.
Copy-editing is the art of preparing typescripts and illustrations for printing and publication. This is the definitive text-book.

Chapter 7
How to write . . . theses

Recommendations for the presentation of theses. BS 4821:1972. London: British Standards Institution. 1972.
If your examining body refers to BS 4821 in its regulations you will need this.

Chapter 8
So you fancy writing a book

Unwin S. *The truth about publishing.* 8th edn. London: George Allen & Unwin. 1976.
Almost a historical classic of publishing (it first appeared in 1926), this is a fascinating insight into the problems, perils and delights of publishing books.

St John Thomas D. *Non-fiction: a guide to writing and publishing.* Newton Abbot: David & Charles. 1970.
This is one of the only two text-books on writing text-books you'll ever come across. Very readable and packed with valuable facts. If you must write a book, read this one first.

Legat M. *An author's guide to publishing.* London: Hale. 1982.
The other book about writing books—or at least, about getting them published. More up-to-date than St John Thomas' book (above) and giving a different viewpoint. Best read them both.

Recommendations for the preparation of indexes. BS 3700:1964. London: British Standards Institution. 1964.
The best (and definitive) guide to preparing an index for your book—or anything else.

Chapter 9
What about word processing?

Varley H, Graham I. *The personal computer handbook.* London: Pan Books. 1983.
An excellent and inexpensive introduction to the world of computers with good sections on word processing and on the selection of hardware. If you are new to this subject then this is probably the best introduction.

Zinsser W. *Writing with a word processor.* New York and London: Harper and Row. 1983.
One man's experience of coming to terms with creative writing on a word processor, this is also the nearest thing to a manual for beginners in the art. Extremely readable and gives an excellent 'feel' of the problems and the satisfactions of using a word processor.

Naiman A. *Introduction to WordStar*™. 2nd edn. Berkeley, Paris and Dusseldorf: Sybex. 1983.
WordStar™ is one of the most sophisticated (and amongst the more expensive) of the word processing software packages. This book gives a good idea of what a word processor can do, with WordStar as an example, and it is very clearly written and well-illustrated (unlike many word processing and computer manuals).

Stananought J. *Word processing. Systems, applications and assignments.* London: McGraw-Hill. 1984.
A detailed and systematic beginner's instruction text in word processing, covering 40 basic functions and written so as to be applicable to any word processing system. Each chapter includes practical exercises. The ideal 'teach yourself' guide.

References

You will find it a very good practice always
to verify your references, sir!

Martin Joseph Routh (1755–1854). In
Burgeon's Memoir of Dr Routh
Quarterly Review, July 1878, vol cxlvi

Introduction

1 Maddox H. *How to study*. London: Pan Books. 1967. p. 163.
2 Day RA. *How to write and publish a scientific paper*. Philadelphia: ISI Press. 1979. p. *iv*.
3 Barrie JM (attributed). In *Familiar Medical Quotations*. 1st edn. Ed. MB Strauss. Boston: Little Brown Co. 1968. p. 671a.

Chapter 1

1 Churchill WS. Election speech at Manchester, 1906. Quoted in Speight R. *Hilaire Belloc*. London: Hollis & Carter. 1957. p. 207.
2 Milne AA. *Winnie-the-Pooh*. Ch. 4.
3 Carroll L. *Through the looking glass*. Ch. 6.
4 Anonymous. Quoted by Gowers E. *The complete plain words*. 2nd edn. Rev. by Fraser B. London: HMSO. 1977. p. 42.
5 Virchow R. *Bull. NY acad. med.*, 1928; **4:** 994.
6 Montgomery WF. Letter to JY Simpson, dated 27 December 1848. Ms in collection of the Royal College of Surgeons of Edinburgh. Quoted in Farr AD, Religious opposition to obstetric anaesthesia: a myth? *Ann. Sci.*, 1983; **4:** 159–77
7 Sheridan RB. *Clio's protest*. See Moore's *Life of Sheridan*, 155.

Chapter 2

1 Kipling R. *The elephant's child*. 1909.

Chapter 3

1 Bradford Hill A. The reasons for writing. *Brit. med. J.*, 1965; **2:** 870.
2 Asher R. Six honest serving men for medical writers. *J. Amer. med. assn.* 1969; **208:** 83–7.
3 Day RA. *How to write and publish a scientific paper*. Philadelphia: ISI Press. 1979. p. 1.

Chapter 4

1 Farr AD. Reference citation. *Earth and Life Science Editing,* 1983; No. 20: 3.

Chapter 5

1 Rich B. 1613. Cited by Barnes SB in *The beginnings of learned journalism, 1665–1730.* Unpublished PhD dissertation. Corell University, Ithaca NY. 1934.

Chapter 6

1 International committee of medical journal editors. Uniform requirements for manuscripts submitted to biomedical journals. *Brit. med. J.* 1982; **284:** 1766–70. Also in *Ann. int. med.* 1982; **96:** 766–71 and *Med. lab. sci.* 1983; **40:** 1–6.

Chapter 8

1 Thomas D St J. *Non-fiction: a guide to writing and publishing.* Newton Abbot: David & Charles. 1970. p. 17.
2 See ref 1. p. 25.
3 Johnson S. In Boswell's *Life,* vol ii. p. 344. 6 April 1775.
4 Race RR, Sanger R. *Blood groups in man.* 2nd edn. 1954. Oxford: Blackwell. p. 272.
5 Legat M. *An authors guide to publishing.* London: Hale. 1982. p. 40.
6 See ref 5. pp. 93–104.

Chapter 9

1 Zinsser W. *Writing with a word processor.* New York and London: Harper and Row. 1983. p. 20.
2 Churchill WS. Speech in London. 1 Oct 1939.
3 Beeton B. Guidelines for author-prepared input for typesetting. In *Scholarly communication around the world.* Washington DC: Society for Scholarly Publishing. 1983. pp. 87–9.
4 Cameron J. (Rapporteur). Editors and text-processing equipment. *J. Res. Comm. Studies.* 1981–2; **3:** 443–7.
5 See Ref 3. p. 89.
6 Watson J. Print on paper publishing via electronic interfaces. In *Scholarly communication around the world.* Washington DC: Society for Scholarly Publishing. 1983. pp. 54–6.
7 *Medical Laboratory Sciences:* statement of editorial policy and full Instructions to Authors. *Med. Lab. Sci.* 1985; **42:** 3.

Acknowledgements

> And so I know not what to call you; but howsoever, I thank you.
>
> Queen Elizabeth I (1533–1603). *Letter to wife of the Archbishop of Canterbury*

Although I had been writing books and articles for many years it was not until I became an editor myself that I started to think critically about the writing process. Since then I have become increasingly involved in trying to teach the elements of communication to scientists, while at the same time imbibing the wisdom of many who are much more experienced at it than I am. I have read widely and listened keenly to others at both national and international conferences and much of what I have learned I have tried to pass on in my own lectures.

In writing this book I am greatly indebted to those who saw the need for a serious study of writing long before I did. I hope that I have not unconsciously plagiarized them or leaned too heavily upon ideas which, although fundamental to the subject, were first learned as lessons expounded by a few pioneers.

In particular it is necessary to refer to six authors, to whose efforts at improving communication every science writer must be either directly or indirectly indebted. Sir Ernest Gowers and Sir Bruce Fraser really destroyed the jargon and cliché-ridden formal English of officialdom in the years following the second world war, while Eric Partridge has probably done as much as anyone to make the common errors and pitfalls of English usage obvious to anyone who cares. In science and medicine Robert Day in America, and Maeve O'Connor and Stephen Lock in Britain, have each blazed a distinctive trail towards the long-overdue improvement of scientists' ability to communicate.

To each of these experts and to all who have in any way taught me anything about writing, I express grateful thanks.

For permission to reproduce certain figures I am indebted to Mrs A. O'Malley and Dr M.D. Penney (Fig. 3), Dr Z. Parvez (Fig. 4), and Messrs P. Robertson, W. Whyte and P.V. Bailey (Fig. 5). These illustrations appeared originally in papers published in *Medical Laboratory Sciences* 1984, 41 (Figs 3 and 4) and 1983, 40 (Fig. 5). Figures 1 and 2 are by the author.

Some of the material in Chapter 7 on the preparation of theses appeared originally in *The Gazette* of the Institute of Medical Laboratory Sciences, 1984; **28:** 114–5.

Index

LINKED